Computer Intensive Methods in Statistics

Computer Intensive Methods in Statistics

Silvelyn Zwanzig
Uppsala University

Behrang Mahjani
Icahn School of Medicine at Mount Sinai

CRC Press
Taylor & Francis Group
Boca Raton London New York

CRC Press is an imprint of the
Taylor & Francis Group, an **informa** business

A CHAPMAN & HALL BOOK

CRC Press
Taylor & Francis Group
6000 Broken Sound Parkway NW, Suite 300
Boca Raton, FL 33487-2742

© 2020 by Taylor & Francis Group, LLC
CRC Press is an imprint of Taylor & Francis Group, an Informa business

No claim to original U.S. Government works

Printed on acid-free paper

International Standard Book Number-13: 978-0-367-19423-9 (Paperback)
978-0-367-19425-3 (Hardback)

This book contains information obtained from authentic and highly regarded sources. Reasonable efforts have been made to publish reliable data and information, but the author and publisher cannot assume responsibility for the validity of all materials or the consequences of their use. The authors and publishers have attempted to trace the copyright holders of all material reproduced in this publication and apologize to copyright holders if permission to publish in this form has not been obtained. If any copyright material has not been acknowledged please write and let us know so we may rectify in any future reprint.

Except as permitted under U.S. Copyright Law, no part of this book may be reprinted, reproduced, transmitted, or utilized in any form by any electronic, mechanical, or other means, now known or hereafter invented, including photocopying, microfilming, and recording, or in any information storage or retrieval system, without written permission from the publishers.

For permission to photocopy or use material electronically from this work, please access www.copyright.com (http://www.copyright.com/) or contact the Copyright Clearance Center, Inc. (CCC), 222 Rosewood Drive, Danvers, MA 01923, 978-750-8400. CCC is a not-for-profit organization that provides licenses and registration for a variety of users. For organizations that have been granted a photocopy license by the CCC, a separate system of payment has been arranged.

Trademark Notice: Product or corporate names may be trademarks or registered trademarks, and are used only for identification and explanation without intent to infringe.

Visit the Taylor & Francis Web site at
http://www.taylorandfrancis.com

and the CRC Press Web site at
http://www.crcpress.com

Contents

Preface ix

Introduction xi

1 **Random Variable Generation** 1
   1.1 Basic Methods 1
        1.1.1 Congruential Generators 5
        1.1.2 The KISS Generator 8
        1.1.3 Beyond Uniform Distributions 9
   1.2 Transformation Methods 11
   1.3 Accept-Reject Methods 16
        1.3.1 Envelope Accept-Reject Methods 20
   1.4 Problems 21

2 **Monte Carlo Methods** 25
   2.1 Independent Monte Carlo Methods 26
        2.1.1 Importance Sampling 30
        2.1.2 The Rule of Thumb for Importance Sampling 32
   2.2 Markov Chain Monte Carlo 35
        2.2.1 Metropolis-Hastings Algorithm 38
        2.2.2 Special MCMC Algorithms 41
        2.2.3 Adaptive MCMC 46
        2.2.4 Perfect Simulation 47
        2.2.5 The Gibbs Sampler 48
   2.3 Approximate Bayesian Computation Methods 52
   2.4 Problems 58

3 **Bootstrap** 61
   3.1 General Principle 61
        3.1.1 Unified Bootstrap Framework 63
        3.1.2 Bootstrap and Monte Carlo 68
        3.1.3 Conditional and Unconditional Distribution 70
   3.2 Basic Bootstrap 72
        3.2.1 Plug-in Principle 73
        3.2.2 Why is Bootstrap Good? 74
        3.2.3 Example where Bootstrap Fails 75
   3.3 Bootstrap Confidence Sets 75

		3.3.1	The Pivotal Method	76
		3.3.2	Bootstrap Pivotal Methods	78
			3.3.2.1 Percentile Bootstrap Confidence Interval	79
			3.3.2.2 Basic Bootstrap Confidence Interval	79
			3.3.2.3 Studentized Bootstrap Confidence Interval	79
		3.3.3	Transformed Bootstrap Confidence Intervals	82
		3.3.4	Prepivoting Confidence Set	83
		3.3.5	BC_a-Confidence Interval	84
	3.4	Bootstrap Hypothesis Tests		86
		3.4.1	Parametric Bootstrap Hypothesis Test	87
		3.4.2	Nonparametric Bootstrap Hypothesis Test	88
		3.4.3	Advanced Bootstrap Hypothesis Tests	90
	3.5	Bootstrap in Regression		91
		3.5.1	Model-Based Bootstrap	91
		3.5.2	Parametric Bootstrap Regression	93
		3.5.3	Casewise Bootstrap in Correlation Model	94
	3.6	Bootstrap for Time Series		97
	3.7	Problems		100

4 Simulation-Based Methods — 105

	4.1	EM Algorithm		106
	4.2	SIMEX		115
	4.3	Variable Selection		123
		4.3.1	F-Backward and F-Forward Procedures	124
		4.3.2	FSR-Forward Procedure	130
		4.3.3	SimSel	132
	4.4	Problems		138

5 Density Estimation — 141

	5.1	Background		141
	5.2	Histogram		143
	5.3	Kernel Density Estimator		145
		5.3.1	Statistical Properties	148
		5.3.2	Bandwidth Selection in Practice	154
	5.4	Nearest Neighbor Estimator		157
	5.5	Orthogonal Series Estimator		157
	5.6	Minimax Convergence Rate		158
	5.7	Problems		161

6 Nonparametric Regression — 163

	6.1	Background		163
	6.2	Kernel Regression Smoothing		166
	6.3	Local Regression		169
	6.4	Classes of Restricted Estimators		173
		6.4.1	Ridge Regression	175

	6.4.2 Lasso	179
6.5	Spline Estimators	181
	6.5.1 Base Splines	182
	6.5.2 Smoothing Splines	187
6.6	Wavelet Estimators	193
	6.6.1 Wavelet Base	193
	6.6.2 Wavelet Smoothing	196
6.7	Choosing the Smoothing Parameter	199
6.8	Bootstrap in Regression	200
6.9	Problems	203

References 207

Index 211

Preface

This textbook arose from the lecture notes of a graduate course on computer intensive statistics offered in the Department of Mathematics at Uppsala University in Sweden.

The first version of the script was written in 2001 when the course was first introduced. Since then, the course has been taught continuously, and the script has been updated and developed over the past 18 years. The course contains 50 lectures, each about 45 minutes long. Several interactive R codes are presented during the lectures. This course requires that students submit four assignments in teams. These assignments may involve solving problems through writing R codes. At the end of the course, the students will write short presentations on a particular research question. As a part of the task, the students will identify an interesting data set to analyze with the methods taught during the course. This is the part of the course that I like the most!

Since its inception in 2001, students from a variety of fields have participated in this course. Many have backgrounds in mathematics or mathematical statistics, but the course has also seen PhD students from other departments (i.e., biology, IT, physics, and pharmaceutics, etc.). PhD students frequently relate their final presentations to their theses.

In 2011, Behrang Mahjani, then a PhD student, came to me and offered collaboration in teaching this course. In the following year, he took over the introduction to R, the random number generation and Monte Carlo methods. In 2016, after earning his PhD, Behrang offered to assist me in finalizing this book. Today, we share the title of "author", and for this, I am thankful. While Behrang and I are the primary authors of this book, we are also joined by a collaborator. This book contains many cartoons, illustrating data material or the main ideas underlying the presented mathematical procedures. The signature on these illustrations, AH, stands for Annina Heinrich, my daughter, a biologist. Based on my explanations and very rough sketches, she drew the cartoons. (My personal favorite pictures are Figures 4.1 and 4.2 about the embedding principle, and Figure 6.2 about smoothing principle).

Last but not least, I would like to thank all of the students who attended this course and gave us feedback at different levels. I am most delighted and proud to hear that my course has been useful when I meet the students long after the course.

Silvelyn Zwanzig
Uppsala, June 2019

Introduction

The promises of modern computational resources are introducing a whole new world of statistical methods. This book is a collection of some of the most important ideas, old and new, that are used in computer intensive methods in statistical analysis. The book also gives insight into the theoretical background of the methods, and how they are related.

Reading the most recently published literature in this field, you might guess that there is no limit to establishing new ideas. However, not every new method is based on a new idea; sometimes, a new method is born from the application of an old idea.

We start by stating a few main principles. They sound obvious, but we should always have them in mind:

- We believe that we can calculate all that we wish, but usually, nothing is calculated exactly. We have no irrational numbers in our calculations, but we have sufficient good approximations.

- Generated random numbers are pseudo-random. Calculation of random numbers and randomness is a contradiction.

- Many methods do not deliver an exact solution after the first step. We have to be sure that the next steps (iterations) improve the solution.

- All methods work under certain assumptions. These assumptions are not usually well stated. Checking the assumptions for real world problems is sometimes complicated. The method might work in practice, but we have to be aware of any underlying assumptions.

- Sometimes it is impossible to solve a problem directly, but you can find a good solution to the approximated problem.

- Rule of thumb: Try to find the analytical solution of the exact problem first, before applying a general approximating iterative procedure.

- Simulation produces additional randomness, which produces a higher variation of the data. Often this effect can be compensated by bigger sample size.

The ideas behind so many different methods are usually not so different. Let us try to sort some of the key ideas.

- Randomization helps. Monte Carlo methods use generated random numbers for approximating an integral. (Chapter 3)

- It is okay to use a wrong random number generator. An additional testing step delivers the right subsample (rejection algorithm). Alternatively, one can correct it by weighting (importance sampling).

- Trial and Error. In the trial step, the numbers are changed a bit, and in the error step, it is tested if the trial was a good step or not. Not every step in a simulation is used to produce the final result (MCMC, ABC).

- Disturbing the data in a controlled way can be used to learn more about them (SIMEX, SIMSEL).

- Change perspectives: Embed the model under consideration in a larger model with latent variables and apply the simpler methods in the large model. The latent variables have to be calculated or simulated cleverly (EM - algorithm, data augmentation, SIMSEL).

- One can expand data set by resampling or simulation from a well-estimated distribution. Then, we can apply the method on the new data sets (Bootstrap).

- Small models can be good fits. Big models are often estimable via a small model. We want to find a good balance between the approximation error by a smaller model and the estimation error in the smaller model (density estimation, bias-variance trade-off, nonparametric regression).

- Many formulas include a free tuning parameter (bandwidth, number of neighbors, degree of depth). The properties of many methods depend essentially on the choice of the tuning parameter. An estimate of the goodness is used for a data-driven choice (kernel density estimate, nonparametric regression).

These ideas can be combined to implement new methods.

In the first chapter, we present the primary methods for generating random variables from any given distribution. The chapter begins with an introduction of methods based on the generalized inverse distribution function. We also introduce congruential generators and KISS generators. Further, we describe transformation methods for generating an arbitrary distributed random number, present the Box-Muller generator, and detail the accept-reject methods.

The second chapter discusses the general principle of Monte Carlo (MC) methods, and Markov Chain Monte Carlo (MCMC) algorithms. We introduce the trial-and-error steps with the historical hard shell gas model. Finally, we demonstrate the approximative Bayesian computation methods (ABC).

Introduction xiii

The third chapter introduces the bootsrap method, a general method that can be beneficial in cases where there is no standard procedure to analyze a model. We describe the relationship between the bootstrap and Monte Carlo methods with the help of Efron's sample median, and then we explore different bootstrap methods for calculating confidence sets and hypothesis testing.

In the fourth chapter, we present methods based on the embedding principle, starting with the expectation-maximization algorithm (EM), and then the relationship between MC and EM in the MCEM algorithm. We introduce the SIMEX method (simulation-extrapolation estimation method), where the data are disturbed in a controlled way. The last section of this chapter presents the variable selection methods, such as the standard F-Backwards and F-Forward procedures. Wu's FSR-forward procedure explores pseudo-variables for estimating the false selecting rate. The SimSel method combines the ideas of the SIMEX and Wu's methods.

The fifth chapter presents methods for density estimation. We have a set of observation from a distribution and we want to estimate an almost arbitrary function with the help of these observations. First, we introduce the histogram, and then the kernel density estimator. Furthermore, we present the nearest neighbor estimators and the orthogonal series estimators.

In the sixth chapter, we explore methods for regression function estimation. We present the relationship between a two-dimensional kernel density estimation and the Nadaraya-Watson estimator. We introduce classes of restricted estimators (e.g., Ridge and Lasso), as methods for estimation of an approximation of the regression function. The base spline estimators and natural spline estimators are explained as least squares estimators in an approximate linear regression model. We introduce the projection idea of wavelet smoothing and the wavelet estimators. Lastly, we explain how to apply the bootstrap confidence band to the estimator of the first derivative of the regression function.

The R code for the book can be found at:
https://github.com/cimstatistics.

Finally, we hope you have fun exploring new ways of analyzing data.

1

Random Variable Generation

In this chapter, we present the primary methods for generating random variables from any given distribution. These methods are frequently used in simulation studies for generating random observations from an estimated distribution.

1.1 Basic Methods

We start by presenting a few of the most fundamental methods for simulating random numbers from a distribution of interest. The *inverse method* is highly applicable to many distributions. This method uses a set of uniformly distributed random numbers on $[0, 1]$ as an input. We will first review some of the principles for generating uniformly distributed random numbers.

> **Definition 1.1** For a distribution function F on $\mathbb{R}$, the **generalized inverse** F^- of F is the function defined by
> $$F^-(u) = \inf\{x,\ F(x) \geq u\} \quad \text{for} \quad 0 \leq u \leq 1.$$

Note that for continuous distribution functions F, the generalized inverse F^- is the inverse function F^{-1}. The function F^- is often called the quantile function and is denoted by Q. As an example, see Figure 1.1.

Example 1.1 Here are four examples where the inverse F^- can be written in an explicit form:
1. *Uniform distribution:* $X \sim \mathsf{U}(a, b)$,

$$F(x) = \frac{x-a}{b-a} I_{[a,\,b]}(x), \qquad F^-(u) = a + (b-a)u.$$

2. *Exponential distribution:* $X \sim \mathsf{Exp}(\lambda)$,

$$F(x) = 1 - e^{-\lambda x}, \quad x > 0, \qquad F^-(u) = -\lambda^{-1}\log(1-u).$$

1

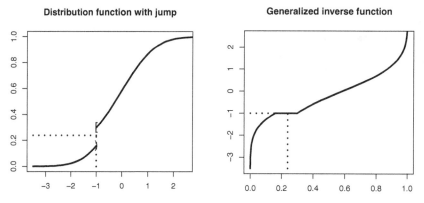

FIGURE 1.1: An example of the generalized inverse distribution function.

3. *Three-point distribution* on $1, 2, 3$ with $p_1 = p_2 = 0.4$, $p_3 = 0.2$: illustrated in Figure 1.2. Then

$$F^-(u) = \begin{cases} 1, & 0 \leq u \leq 0.4, \\ 2, & 0.4 < u \leq 0.8, \\ 3, & 0.8 < u \leq 1. \end{cases}$$

4. *Cauchy distribution*: $X \sim \mathsf{C}(0, 1)$, where

$$f(x) = \frac{1}{\pi(1 + x^2)}, \quad -\infty < x < \infty.$$

Then

$$F(x) = \frac{1}{2} + \frac{1}{\pi}\arctan x, \quad F^-(u) = \tan(\pi(u - 1/2)).$$

□

The following lemma demonstrates that a uniform distributed random variable transformed by the generalized inverse has the desired distribution.

Lemma 1.1 *If* $U \sim U[0, 1]$, *then the random variable* $F^-(U)$ *has the distribution* F.

Proof: We have to show

$$F^-(u) \leq x \iff u \leq F(x), \tag{1.1}$$

because under (1.1) it holds

$$\mathsf{P}\left(F^-(u) \leq x\right) = \mathsf{P}(u \leq F(x)) = F(x).$$

1.1 Basic Methods

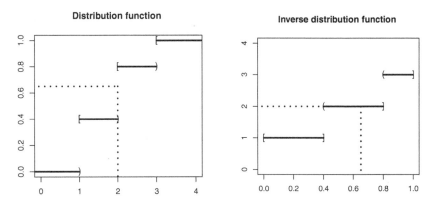

FIGURE 1.2: Generalized inverse distribution function of the three-point distribution in Example 1.1.

It remains to show (1.1). First: $\Longrightarrow$ Let $F^-(u) \leq x$. The distribution function F is monotone, thus
$$F\left(F^-(u)\right) \leq F(x).$$
Using the definition of F^-
$$F\left(F^-(u)\right) = F(\inf\{x, F(x) \geq u\}) \geq u,$$
thus, $u \leq F\left(F^-(u)\right) \leq F(x)$.

The next step is $\Longleftarrow$: Let $u \leq F(x)$. The inverse distribution function F^- is monotone, thus
$$F^-(u) \leq F^-(F(x)).$$
If we set $h = F(x)$ and apply the definition of the general inverse distribution function, we get
$$F^-(F(x)) = F^-(h) = \inf\{z, \ F(z) \geq h\}.$$
Using $h = F(x)$, we obtain $x \in \{z, F(z) \geq h\}$, and
$$x \geq F^-(F(x)) \geq F^-(u).$$

$\square$

The main message of Lemma 1.1 can be stated as:

> All we need is a good generator for $\mathsf{U}[0,1]$ distributed random numbers!

Example 1.2 (Continuation of Example 1.1) Assuming we have $X \sim \mathsf{U}(0,1)$, we can simulate some of the standard distributions.

1. *Uniform distribution:* $X \sim \mathsf{U}(a,b)$,
$$X = a + (b-a)U, \quad U \sim \mathsf{U}(0,1).$$

2. *Exponential distribution:* $X \sim \mathsf{Exp}(\lambda)$,
$$X = -\frac{1}{\lambda}\log(1-U) \sim \mathsf{Exp}(\lambda) \quad \text{and} \quad -\frac{1}{\lambda}\log(U) \sim \mathsf{Exp}(\lambda).$$

$\square$

We now focus on algorithms for generating i.i.d. numbers $u_1, \ldots, u_n$ from $\mathsf{U}(0,1)$.

> **Definition 1.2** A **uniform pseudo-random number generator** is an algorithm which, starting from an initial value u_0 and a transformation D, produces a sequence with $u_i = D^i(u_0)$ in $[0,1]$. For all n, $(u_1, \ldots, u_n)$ reproduce the behavior of an i.i.d. sample of uniform random variables when compared through a set of tests.

Keep in mind that the pseudo-random numbers are not truly random. They are fake! In other words:

Deterministic systems are used to imitate a random phenomenon!

There are several properties that should be verified to check if a generated sample has the behavior of an i.i.d. These properties are necessary but not sufficient:

1.1 Basic Methods

- Output numbers are almost uniformly distributed.
- Output numbers are independent.
- The period between two identical numbers is sufficiently long.
- It should not be possible to predict the next member in the sequence.

Some commonly used tests are Kolmogorov–Smirnov tests, chi-square tests, run tests, permutation tests, and autocorrelation tests. In addition, scatter plots, $(u_i, u_{(i-k)})$, are usually helpful to detect patterns in a sequence of pseudo-random numbers. As an example, see Figure 1.3.

1.1.1 Congruential Generators

Congruential generators constitute a common methodology for generating samples, with roots tracing back to the 1950s. The mathematical background is from number theory. We start with the definition of *period* in this context.

> **Definition 1.3** The **period** T_0 of a generator is the smallest integer T such that
> $$u_{i+T} = u_i \quad \text{for every } i.$$

> **Definition 1.4** A **congruential generator (a,b,M)** on $\{0, \ldots, M-1\}$ is defined by the function
> $$D(x) = (ax + b) \mod M,$$
> where a and b are integers chosen from $\{0, 1, \ldots, M-1\}$.

The expression $y = z \mod M$, where y and z are integer-valued expressions, means that there is an integer k such that $kM = y - z$. The coefficient a can be interpreted as a "multiplier", and the coefficient b as an "increment". This method is referred to as mixed-congruential, but when $b = 0$, it is called a multiplicative-congruential. The starting value x_0 is called a *seed*.

With this generator, the period should not be greater than M. One can use the following transformation to deliver a generator over $[0, 1]$:

$$\widetilde{D}(x) = \frac{D(x)}{M} \mod 1.$$

Example 1.3 1. The first published example, according to Ripley (1987), used $a = 2^7$, $b = 1$, $M = 2^{35}$, hence the generator $(128, 1, 2^{35})$.

2. The generator (69069, 1327217885, 2^{32}), i.e.,

$$x_i = (69069 x_{i-1} + 1327217885) \bmod 2^{32},$$

has been successfully implemented in Fortran 90.

3. Another interesting example is the (now infamous) generator RANDU, with $M = 2^{31}$, $a = 2^{16}+3$, $b = 0$. There is a correlation between the points, because for an integer c

$$x_i = 6 x_{i-1} - 9 x_{i-2} + c 2^{31}.$$

□

R Code 1.1.1. Congruential generator

```
random <- function(seed,n,M,a,c)
{
rand <- rep(NA,n)
rand[1] <- seed
for(i in 2:n){
  rand[i] <- (a*rand[i-1]+c)%%M
  }
return(rand/M)
}
```

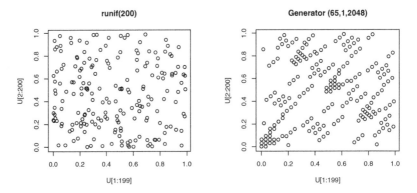

FIGURE 1.3: Scatter plot for detecting patterns.

1.1 Basic Methods

Exercise 1.1 Use R and plot (u_i, u_{i+1}) with $u_i = x_i/M$, $M = 2^{11}$, $x_0 = 0$, and
$$x_i = ax_{i-1} + c \bmod M,$$
for following cases:

(a) Plot 2^{11} points with $a = 65, c = 1$.

(b) Plot 512 points for these combinations: $(a = 1365, c = 1)$, $(a = 1229, c = 1)$, $(a = 157, c = 1)$, $(a = 45, c = 0)$, $(a = 43, c = 0)$.

Theorem 1.1 *The congruential generator specified by (a, b, M) has period M if and only if*

(i) $\gcd(b, M) = 1$, i.e., b and M are relatively prime.

(ii) $p \mid (a - 1)$ for every prime p such that $p \mid M$.

(iii) $4 \mid (a - 1)$ if $4 \mid M$.

Note: gcd stands for the greatest common divisor. $m|n$ means m is a divisor of n. For proof of Theorem 1.1, and more results with respect to congruential generators, see Kennedy and Gentle (1980) and Ripley (1987).

The following is an interesting property of the generators when M is large.

Let $(x_1, \ldots, x_M)$ be generated by a congruential generator with full period. The outcomes will then $(x_1, \ldots, x_M) = (0, 1, \ldots, M-1)$ all have the probability $1/M$. For the related random variable X, we find

$$\mathsf{E}[X] = \frac{1}{M} \sum_{j=0}^{M-1} x_j = \frac{1}{M} \sum_{j=0}^{M-1} j = \frac{M-1}{2},$$

and further

$$\mathsf{Var}[X] = \frac{1}{M} \sum_{j=0}^{M-1} x_j^2 - \left(\frac{M-1}{2}\right)^2 = \frac{1}{M} \sum_{j=0}^{M-1} j^2 - \left(\frac{M-1}{2}\right)^2 = \frac{M^2 - 1}{12}.$$

Thus, for $U = X/M$,

$$\mathsf{E}[U] = \frac{1}{2} - \frac{1}{2M}, \quad \mathsf{Var}[U] = \frac{1}{12} - \frac{1}{12M^2}.$$

In other words, for large M, $U = X/M$ has asymptotically the moments of $\mathsf{U}(0, 1)$.

1.1.2 The KISS Generator

The following generator, originally presented by Marsaglia and Zaman (1993) (see Robert and Casella (1999) for further discussion) has been implemented in GNU Fortran. The name, KISS, is an acronym for "**K**eep **I**t **S**imple, **S**tupid!". This algorithm is based on a shift-register technique. Before presenting the shift-register generator, we define the binary representation of a number. The binary representation of x_n is a vector of binary coordinates $e_n = (e_{n,0}, \ldots e_{n,k-1})^T$, $e_{n,j} \in \{0,1\}$, and $x_n = \sum_{j=0}^{k-1} e_{n,j} 2^j$.

> **Definition 1.5** The **shift-register generator** for a given $k \times k$ matrix T, with entries 0 or 1, is given by the transformation
> $$e_{n+1} = T e_n.$$

The generators used by KISS are based on the $k \times k$ matrices

$$T_L = \begin{pmatrix} 1 & 1 & 0 & \cdots & 0 \\ 0 & 1 & 1 & \ddots & \vdots \\ \vdots & 0 & \ddots & 1 & \vdots \\ \vdots & \ddots & 0 & 1 & 1 \\ 0 & \cdots & \cdots & 0 & 1 \end{pmatrix}, \quad T_R = \begin{pmatrix} 1 & 0 & \cdots & \cdots & 0 \\ 1 & 1 & 0 & \ddots & \vdots \\ \vdots & 1 & \ddots & 0 & \vdots \\ \vdots & \ddots & 1 & 1 & 0 \\ 0 & \cdots & 0 & 1 & 1 \end{pmatrix}.$$

These matrices are related to the right and left shift matrices, R and L,

$$R \begin{pmatrix} e_1 \\ e_2 \\ \vdots \\ e_k \end{pmatrix} = \begin{pmatrix} 0 \\ e_1 \\ \vdots \\ e_{k-1} \end{pmatrix}, \quad L \begin{pmatrix} e_1 \\ e_2 \\ \vdots \\ e_k \end{pmatrix} = \begin{pmatrix} e_2 \\ \vdots \\ e_k \\ 0 \end{pmatrix},$$

since $T_R = (I + R)$ and $T_L = (I + L)$, where I is the identity matrix.

> **Algorithm 1.1 KISS Generator**
>
> 1. Define Congruential generator as
> $$I_{n+1} = (69060 \times I_n + 23606797) \mod 2^{32}.$$
>
> 2. Define two shift generators as
> $$\begin{aligned} J_{n+1} &= (I + L^{15})(I + R^{17}) J_n \mod 2^{32}, \\ K_{n+1} &= (I + L^{13})(I + R^{18}) K_n \mod 2^{31}. \end{aligned}$$

1.1 Basic Methods

> 3. Convert the binary coordinate vectors J_{n+1} and K_{n+1} to decimal numbers j_{n+1} and k_{n+1}. Combine steps 1 and 2:
>
> $$x_{n+1} = (I_{n+1} + j_{n+1} + k_{n+1}) \bmod 2^{32}.$$

Note that some care must be taken when using the KISS generator on [0,1], as it implies dividing by the largest integer available on the computer! The KISS generator has been successfully tested. For more details, see pp. 39-43 of Robert and Casella (1999).

1.1.3 Beyond Uniform Distributions

It can be computationally expensive to calculate the generalized inverse, the method based on Lemma 1.1. Particular methods have been developed to simulate samples from some distributions. For example, it is fairly easy to simulate samples from exponential, double exponential (Gumbel), and Weibull distributions. In some cases, one can use approximations of F^-.

An intuitive example is generating random numbers from a discrete distribution. Consider an arbitrary discrete distribution with the probability function

$$\begin{array}{cccc} i: & 1 & \ldots & K, \\ p_i: & p_1 & \ldots & p_K. \end{array}$$

Then

$$F(k) = \sum_{i=1}^{k} p_i,$$

and

$$F^-(u) = i - 1 \quad \text{for } u \text{ with } F(i-1) < u \leq F(i).$$

The following example is a six-point distribution where the probabilities $p_i, i = 1, \ldots, 6$, are additionally parameterized.

Example 1.4 (The Hardy-Weinberg principle) The Hardy-Weinberg principle characterizes the distributions of genotype frequencies in a given population. Consider a large population where there are three possible alleles S, I, F at one locus, resulting in six genotypes

$$\text{SS, II, FF, SI, SF, IF.}$$

Let θ_1, θ_2 and θ_3 denote the probabilities of S, I and F, respectively, with $\theta_1 + \theta_2 + \theta_3 = 1$. The Hardy-Weinberg principle assigns the following probabilities for the genotypes:

Genotype	1	2	3	4	5	6
Genotype	SS	II	FF	SI	SF	IF
Probability	θ_1^2	θ_2^2	θ_3^2	$2\theta_1\theta_2$	$2\theta_1\theta_3$	$2\theta_2\theta_3$

In other words, $p_1 = \theta_1^2$ and so forth. □

The following algorithm is based on Lemma 1.1.

Algorithm 1.2 Generation from a Discrete Distribution

For i from 1 to n:

1. Generate a random number u from $U \sim \mathsf{U}(0,1)$.
2. If $\sum_{l=1}^{i} p_l < u$ do $i + 1$.
3. If $\sum_{l=1}^{i} p_l \geq u$ return $X = i$.

This algorithm implies that for $X = i$, we will have to do i comparisons, which can be highly time-consuming. In order to decrease the computational costs of this algorithm, one possibility is to apply a search algorithm such as the binary search tree.

Exercise 1.2 Consider Example 1.4 and assume that $\theta_1 = 0.75$, $\theta_2 = 0.15$, $\theta_3 = 0.10$. Write R code that simulates a sample of size 1000.

Several algorithms have been proposed over the years for simulation of the normal distributed random variables.

Example 1.5 (Normal distribution) The distribution function $\Phi(x)$ for a random variable $X \sim \mathsf{N}(0,1)$ cannot be expressed explicitly. One can use the approximation

$$\Phi(x) \simeq 1 - \varphi(x)\left[b_1 t + b_2 t^2 + b_3 t^3 + b_4 t^4 + b_5 t^5\right], \; x > 0,$$

where $\varphi(x)$ denotes the density for a standard normal distribution, and

$$t = \frac{1}{1+px}, \; p = 0.2316419,$$
$$b_1 = 0.31938, \; b_2 = -0.35656, \; b_3 = 1.73148, \; b_4 = -1.82125, \; b_5 = 1.33027.$$

1.2 Transformation Methods

Similarly, for the inverse we have

$$\Phi^{-1}(\alpha) \simeq t - \frac{a_0 + a_1 t}{1 + b_1 t + b_2 t^2},$$

where

$t^2 = \log(\alpha^{-2})$, $a_0 = 2.30753$, $a_1 = 0.27061$, $b_1 = 0.99229$, $b_2 = 0.04481$.

These two approximations are exact up to an error of order 10^{-8}. For more details, see Abramowitz and Stegun (1964). Another approach is to calculate the inversion of $\Phi(u) = u$ by a numerical procedure, see Devroye (1985).

1.2 Transformation Methods

Transformation methods are useful when a distribution F is linked, in a relatively simple way, to another distribution that is easy to simulate from.

> **Theorem 1.2** *Let X have a distribution function F and let $h : \mathbb{R} \to B \subseteq \mathbb{R}$ strictly increasing. Then $h(X)$ is a random variable with distribution function $F(h^{-1}(x))$.*
>
> *If F has a density f and h^{-1} is absolutely continuous, then $h(x)$ has a density $(h^{-1})'(x) f(h(x))$.*

For a proof, see Gut (1991) p. 23.

Example 1.6 (Exponential distribution) Exponential random numbers are obtained in a straightforward way; if $U \sim \mathsf{U}(0,1)$, then $-\log(U)/\lambda \sim \mathsf{Exp}(\lambda)$.

For i.i.d. variables $U_i \sim \mathsf{U}(0,1)$, one can show the following identities

$$Y = -b \sum_{i=1}^{a} \log(U_i) \sim \Gamma(a,b),$$

$$Y = -2 \sum_{i=1}^{n} \log(U_i) \sim \chi_{2n}^2, \tag{1.2}$$

and

$$Y = \frac{\sum_{i=1}^{a} \log(U_i)}{\sum_{i=1}^{a+b} \log(U_i)} \sim \mathsf{Beta}(a,b),$$

which can be used for simulation purposes. $\square$

Example 1.7 (Mixture distribution) A random variable, X, has density $f(x)$ where
$$f(x) = \alpha\, f_1(x) + (1-\alpha)\, f_2(x),$$
and $f_1(x)$ and $f_2(x)$ are densities of the distributions F_1 and F_2, and $\alpha \in [0,1]$. X can be presented as

$$X = ZX_1 + (1-Z)X_2,\ X_1 \sim F_1, X_2 \sim F_2, Z \sim \mathsf{Ber}(\alpha), \text{independently.}$$

We generate independent samples $x_{1,1}, \ldots, x_{1,n}$ from F_1, $x_{2,1}, \ldots, x_{2,n}$ from F_2, and $z_1, \ldots, z_n$ from $\mathsf{Ber}(\alpha)$. Then, we calculate X's as

$$x_i = z_i x_{1,i} + (1-z_i) x_{2,i},\ i = 1, \ldots, n.$$

As an example, see Figure 1.4. □

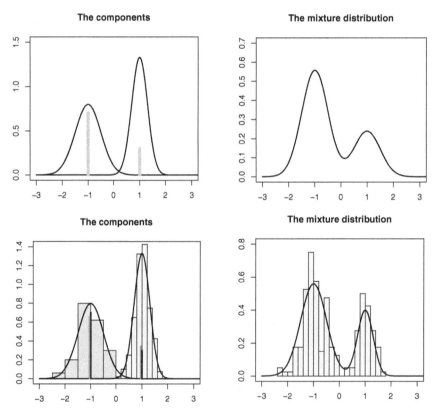

FIGURE 1.4: Mixture distribution in Example 1.7. Vertical lines are potential α and $1-\alpha$.

1.2 Transformation Methods

R Code 1.2.2. Generate from a mixture distribution

```
x1 <- rnorm(200,-1,0.5) ### 200 random numbers from N(-1,0.25)
x2 <- rnorm(200,1,0.3)  ### 200 random numbers from N(1,0.09)
bin <- rbinom(200,1,0.7) ### 200 random numbers from Ber(0.7)
z <- bin*x1+(1-bin)*x2 ### mixture
```

Example 1.8 (Normal distribution) Let (X_1, X_2) be a bivariate standard normal distribution. Let r and θ be the polar coordinates of (X_1, X_2), then

$$r^2 = X_1^2 + X_2^2 \sim \chi^2(2).$$

The distribution of (X_1, X_2) is rotation invariant, see Figure 1.5. Hence $\theta \sim \mathsf{U}(0, 2\pi)$. Furthermore, the distance r^2 and the angle θ are independent random variables. From (1.2), we know that $-2\log(U) \sim \chi^2(2)$ for $U \sim \mathsf{U}(0,1)$, and $2\pi U \sim \mathsf{U}(0, 2\pi)$ for $U \sim \mathsf{U}(0,1)$. If (U_1, U_2) are i.i.d. $\mathsf{U}(0,1)$, the variables X_1, X_2 are i.i.d. standard normally distributed and can be defined by

$$X_1 = \sqrt{-2\log(U_1)}\cos(2\pi U_2), \quad X_2 = \sqrt{-2\log(U_1)}\sin(2\pi U_2). \tag{1.3}$$

□

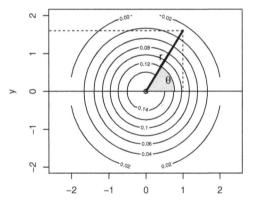

FIGURE 1.5: Polar coordinates in a contour plot of bivariate normal distribution.

The classical method of generating normal random numbers using (1.3) was proposed by Box and Muller (1958).

Algorithm 1.3 The Box-Muller Algorithm

1. Generate (U_1, U_2) i.i.d. from $U(0,1)$.
2. Define
$$x_1 = \sqrt{-2\log(U_1)}\cos(2\pi U_2), \quad x_2 = \sqrt{-2\log(U_1)}\sin(2\pi U_2).$$
3. Take x_1 and x_2 as two independent draws from $N(0,1)$.

Remark 1.1 The properties of Algorithm 1.3 depend on the random number generator for (U_1, U_2). In Ripley (1987), examples are given that less successful uniform number generators can produce patterns that are appealing to the eye, but whose statistical properties are not wanted. For example: $x_i = (65x_{i-1} + 1)\mathrm{mod}\ 2048$, see Ripley (1987), p. 5. Examples are given in Figure 1.6 and Figure 1.7.

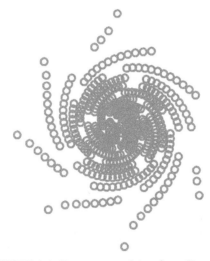

FIGURE 1.6: Patterns resulting from Box-Muller.

R Code 1.2.3. Box-Muller algorithm

```
picture <- function(seed1,seed2,n,N,M,a,b)
{
  u1 <- random(seed1,N,M,a,b)
```

1.2 Transformation Methods

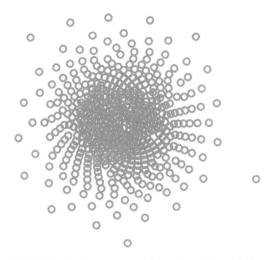

FIGURE 1.7: Patterns resulting from Box-Muller.

```
  u2 <- random(seed2,N,M,a,b)
  x1 <- sqrt(-2*log(u1))*cos(2*pi*u2)
  x2 <- sqrt(-2*log(u1))*sin(2*pi*u2)
  plot(x1[n:N],x2[n:N],"p",pch=21,asp=1,col=1,
  axes=F,ylab="",xlab="", ylim=c(-3,3),xlim=c(-3,3))
}
### asp=1 is important for a symmetric picture!
picture(2,4,10,500,2048,65,1)### rosette
picture(0,2,100,500,2048,129,1)### spiral
```

The following example illustrates that observations from a discrete distribution can be simulated based on random numbers from a continuous distribution.

Example 1.9 (Poisson distribution) As in Example 1.6, one can easily generate exponential random numbers. If $N \sim Po(\lambda)$ and $X_i \sim Exp(\lambda)$, then Poisson random numbers can be generated using

$$\mathsf{P}_\lambda (N = k) = \mathsf{P}_\lambda(X_1 + X_2 + \cdots + X_k \leq 1 < X_1 + X_2 + \cdots + X_{k+1}).$$

□

1.3 Accept-Reject Methods

The main idea in accept-reject methods is to generate random numbers from a "wrong" distribution and to decide with an additional random test which of the wrong generated random numbers fits the desired distribution. Assume we want to generate a sample with distribution $p(x)$ (the *target density*). Consider a distribution $g(x)$ from which it is easy to draw a sample (the *instrumental density*, or *trial distribution*). Suppose that

$$p(x) \leq Mg(x) \text{ for all } x.$$

Rejection Sampling (von Neumann (1951)):

1. Draw X from $g(.)$ and compute the ratio

$$r = \frac{p(x)}{Mg(x)}.$$

2. Flip a coin with success probability r:

   If it is heads, accept and return the value of X; otherwise, reject the value of X and go back to 1.

We formulate this randomized sampling strategy as the following algorithm.

Algorithm 1.4 The Rejection Algorithm

Given a current state $x^{(t)}$.

   1. Draw $X = x$ from $g(.)$ and compute the ratio $r(x) = \frac{p(x)}{Mg(x)}$.
   2. Draw $U \sim \mathsf{U}(0,1)$

$$x^{(t+1)} = \begin{cases} x & \text{if } U \leq r(x), \\ \text{new trial} & \text{otherwise.} \end{cases}$$

The generated sample $\left(x^{(1)}, \ldots, x^{(n)}\right)$ is an i.i.d. sample from the target distribution.

Why does the rejection algorithm work?

The updated variable has the distribution $p(x \mid I = 1)$, where $I = 1$ if x is accepted and $I = 0$ otherwise. Therefore, we obtain $\mathsf{P}(I = 1 \mid X = x) = r(x)$.

1.3 Accept-Reject Methods

From a version of the law of total probability, we have

$$P(I=1) = \int P(I=1 \mid X=x) g(x)\, dx = \int \frac{p(x)}{Mg(x)} g(x)\, dx = \frac{1}{M},$$

and thus

$$p(x \mid I=1) = \frac{P(I=1, X=x)}{P(I=1)} = \frac{p(x)}{Mg(x)} g(x)\, M = p(x).$$

Example 1.10 (Beta distribution) A random variable $X \sim \text{Beta}(a,b)$ has the density function

$$f(x) = cx^a(1-x)^b, \quad 0 \le x \le 1,$$

where $a > 0$, $b > 0$, and c is the normalization constant that can be expressed in terms of the gamma function:

$$c = \frac{\Gamma(a+b)}{\Gamma(a)\Gamma(b)}.$$

As an example, see Figure 1.8.

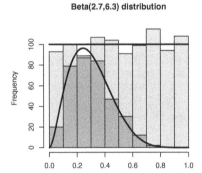
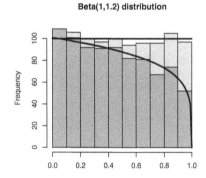

FIGURE 1.8: Sample of instrumental distribution and subsample from the trial distribution and the respected rescaled densities.

□

R Code 1.3.4. Generation from Beta(a,b)

```
Nsim <- 1000
a <- 2.7
b <- 6.3
```

```
M <- 2.67
u <- runif(Nsim,max=M)### uniformly distributed over (0,M)
y <- runif(Nsim) ### instrumental sample
x <- y[u<dbeta(y,a,b)] ### accepted subsample
length(x) ### random (!) sample size of the generated  sample
```

In the rejection algorithm, because the acceptance probability is $\frac{1}{M}$, the constant M should be as small as possible. In the following example, we study this aspect closer.

Example 1.11 (Normal generation from a Cauchy) The Cauchy distribution has heavier tails than the standard normal distribution, (see Figure 1.9); thus, it is possible to find a constant M such that

$$\frac{1}{\sqrt{2\pi}}e^{-x^2/2} \leq M\frac{1}{\pi}\frac{1}{1+x^2}, \quad \text{for all } x.$$

Both densities are symmetric around 0, the optimal M_{opt} allows that the curves touch at two points:

$$\frac{M_{\text{opt}}g'(x)}{p'(x)} = 1 \text{ and } \frac{M_{\text{opt}}g(x)}{p(x)} = 1.$$

That implies

$$\frac{g'(x)}{g(x)} = \frac{p'(x)}{p(x)}.$$

In our case, we have $p'(x) = -xp(x)$ and $g'(x) = -2x(1+x^2)^{-1}g(x)$. Thus, the points are $x = -1$, $x = 1$, and

$$M_{\text{opt}} = \frac{p(1)}{g(1)} = \sqrt{2\pi}e^{-1/2} = 1.520346901,$$

see Figure 1.9. □

In contrast to the R code for the generation of Beta(a,b)-variables, the sample size N is fixed, and the number of simulations is random in the following R-code.

R Code 1.3.5. Normal generation from a Cauchy

```
rand.reject <- function(N,M)
{
rand <- rep(NA,N)
for(i in 1:N){
  L<-TRUE
  while(L){
```

1.3 Accept-Reject Methods

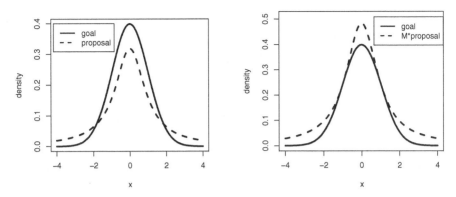

FIGURE 1.9: Determining the constant M for the rejection algorithm.

```
        rand[i] <- rcauchy(1)
        r <- dnorm(rand[i])/(M*dcauchy(rand[i]))
        if(runif(1)<r){
            L <- FALSE}
      }
   }
   return(rand)
}
R <- rand.reject(100,1.7) ### Here, the sample size is 100.
shapiro.test(R) ### Test for normal distribution
```

In some cases, calculating the normalizing constant of the target density $p(x)$ is complicated or time consuming, for instance the $\Gamma(a,b)$-distribution,

$$p(x) = \frac{b^a}{\Gamma(a)} x^{a-1} \exp(-bx), \quad x > 0.$$

Especially in Bayesian, analysis of the posterior distribution is given by its kernel,

$$p(\theta|x) \propto p(\theta)\, l(\theta;x).$$

The rejection algorithm is beneficial for generating random numbers from the posterior $p(\theta|X)$, since one can use the rejection algorithm with an unknown normalizing constant. Suppose the target density $p(x)$ has the kernel $f(x)$

$$p(x) = cf(x) \text{ and } f(x) \leq Mg(x).$$

Set

$$r(x) = \frac{f(x)}{Mg(x)}.$$

Analogously we get

$$P(I=1) = \int r(x)g(x)dx = \frac{1}{M}\int f(x)dx = \frac{1}{M}\int \frac{1}{c}p(x)dx = \frac{1}{Mc},$$

and

$$P(x|I=1) = Mc\, r(x)g(x) = Mc\frac{f(x)}{Mg(x)}g(x) = cf(x) = p(x).$$

One criticism to the accept-reject method is that it generates "useless" simulations when rejecting. An alternative approach is importance sampling. For more details, see Chapter 2, or Robert and Casella (1999) and Lange (1999).

1.3.1 Envelope Accept-Reject Methods

This method is also called squeezing principle. It is useful in situations where the calculation of the target probability is time-consuming. The test step of the rejecting algorithm is split into a pretest and a test step.

Algorithm 1.5 The Squeezing Algorithm
Suppose there exist a density g (instrumental), a function g_l, and a constant M such that

$$g_l(x) \leq p(x) \leq M\, g(x).$$

1. Generate $X \sim g(x)$, $U \sim \mathsf{U}(0,1)$.
2. Accept X if $U \leq \frac{g_l(x)}{Mg(x)}$.
3. Otherwise, accept X if $U \leq \frac{p(x)}{Mg(x)}$.

Algorithm 1.5 produces random variables that are distributed according to $p(x)$.

Example 1.12 (Continuation of Example 1.11) Because

$$\exp(-\frac{x^2}{2}) \geq 1 - \frac{x^2}{2}$$

we can apply the squeezing algorithm with $M = 1.52$ and

$$g_l(x) = \frac{1}{\sqrt{2\pi}}\left(1 - \frac{x^2}{2}\right),\ p(x) = \frac{1}{\sqrt{2\pi}}\exp(-\frac{x^2}{2}),\ g(x) = \frac{1}{\pi}\frac{1}{1+x^2}.$$

See Figure 1.10. □

For further reading, see Chapter 2 of Robert and Casella (1999).

1.4 Problems

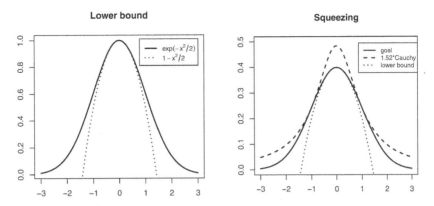

FIGURE 1.10: Enveloping the target density by the lifted up instrumental density and a lower bound in Example 1.12.

1.4 Problems

1. (a) Generate 4 samples of a normal distribution with expectation 3 and variance 2 of size: 10, 100, 1000, 10 000.
   (b) Test that the first and second samples are normally distributed (ks, shapiro, chisq).
   (c) Plot the histogram of all samples.
   (d) Explain how the pseudo-random normal distributed r.v. are generated in R.

2. Consider the continuous distribution with density
$$p(x) = \frac{1}{2}\cos x \, , \, -\frac{\pi}{2} < x < \frac{\pi}{2}. \tag{1.4}$$

   (a) Plot the distribution function $F(x)$.
   (b) Find the inverse distribution function $F^{-1}(u)$.
   (c) Plot the inverse distribution function $F^{-1}(u)$.
   (d) Assume that there is a random number generator for uniform distribution over the interval [0,1]. Write an algorithm to generate a random number from (1.4).
   (e) Write an R code for (d).

3. Consider the discrete distribution
$$P(X=k) = p_k = \binom{3}{k}\left(\frac{1}{3}\right)^k\left(\frac{2}{3}\right)^{3-k}, \, k=0,\ldots,3. \tag{1.5}$$

   (a) Plot the distribution function $F(x)$.

(b) Find the inverse distribution function $F^-(u)$.

(c) Plot the inverse distribution function $F^-(u)$.

(d) Assume that there is a random number generator for a uniform distribution over the interval [0,1]. Write an algorithm to generate random numbers from (1.5).

4. Consider the following R code for generating random numbers:

```
random <- function(n){
    u <- runif(n)
    x <- rep(0,n)
    for(i in 1:n){
        if (u[i]<=0.5){x[i]=-1}
        else if(u[i]<=0.6){x[i]=0}
        else if(u[i]<=0.8){x[i]=1}
        else {x[i]=2}
    }
    return(x)
}
random(100)
```

(a) What is the distribution of the generated sample?

(b) Plot the distribution function $F(x)$.

(c) Determine the inverse distribution function $F^-(u)$.

(d) Plot the inverse distribution function $F^-(u)$.

5. Consider the Hardy-Weinberg model with 6 genotypes. Let $\theta_1, \theta_2, \theta_3$ denote the probabilities of the alleles S, I, F with $\theta_1 + \theta_2 + \theta_3 = 1$. The Hardy-Weinberg principle assigns the following probabilities to the six genotypes:

Genotype	SS	II	FF	SI	SF	IF
Probability	θ_1^2	θ_2^2	θ_3^2	$2\theta_1\theta_2$	$2\theta_1\theta_3$	$2\theta_2\theta_3$

(1.6)

(a) Use the generator runif in R and write an R code for generating random variables following the distribution (1.6) with $\theta_1 = \theta_2 = \frac{1}{3}$.

(b) Apply the Chi-squared test on a sample of size 100, generated by the code in (a), in order to check your simulated sample.

6. Write your own congruential generator (a,b,M) in R.

(a) Find a suitable combination of a,b,M.

(b) Apply the Kolmogorov Smirnov test to a generated sample of size 10 and 100.

(c) Make a scatter plot of (U_i, U_{i-1}) for a sufficiently large sample.

1.4 Problems

(d) Find an unreasonable combination of (a,b,M) where the generated sample with this unreasonable generator passes the Kolmogorov Smirnov test.

7. Use your generator from problem 6 to generate a sample of size 100 from the following distributions:

    (a) An exponential distribution with $\lambda = 2$. Plot the histogram. Apply the ks.test.
    (b) A Cauchy C(0,1) distribution. Plot the histogram. Apply the ks.test.

8. Construct a deterministic sequence which passes the Kolmogorov Smirnov test for $H_0 : U[0,1]$.

9. Assume that there is a good generator for $U[0,1]$ random variables. The aim is to generate an i.i.d. sample from $N_3(\mu, \Sigma)$ with

$$\mu = \begin{bmatrix} 1 \\ 2 \\ 3 \end{bmatrix} \quad \Sigma = \begin{bmatrix} 5 & 2 & 4 \\ 2 & 2 & 2 \\ 4 & 2 & 6 \end{bmatrix}.$$

   (a) Introduce polar coordinates for $X \sim N_2(0, I)$.
   (b) Which distribution has the polar coordinates? Explain it geometrically.
   (c) Describe the main steps of the Box Muller Algorithm.
   (d) Write the main steps for an algorithm generating $N_3(\mu, \Sigma)$ random variables.

10. Given the following R code for generating a random sample:

```
random <- function(N,M){
    rand <- rep(NA,N)
    for(i in 1:N){
        L < -TRUE
        while(L){
        rand[i] <- rcauchy(1)
        r <- dt(rand[i],8)/(M*dcauchy(rand[i]))
        if(runif(1)<r){L <- FALSE}
        }
    }
    return(rand)
}
```

   (a) What is the distribution of the generated sample?
   (b) Which trial distribution was used?
   (c) Write the steps of the algorithm.
   (d) Which choice of $M \in \{1, 1.5, 2\}$ may be possible? Discuss the effect of the bound M.

11. Assume a kernel density estimator $\widehat{f}(x)$ is given with two local maxima at -1 and 2. Suppose there are random number generators (runif) for the uniform distribution, and for the normal distribution (rnorm) $N(0,1)$. The goal is to generate samples from $\widehat{f}(x)$.

   (a) Propose an instrumental density $g_1(x)$ for a rejection algorithm using the random number generator (rnorm) only. Propose a constant M_1 such that $\widehat{f}(x) \leq M_1 g_1(x)$. Describe the principle of choosing a convenient bound M_1. Plot $\widehat{f}(x)$ and $M_1 g_1(x)$.

   (b) Determine an instrumental density $g_2(x)$ for a rejection algorithm using the random number generators (rnorm) and (runif). Propose a constant M_2, such that $\widehat{f}(x) \leq M_2 g_2(x)$. Describe the principle of choosing a convenient bound M_2. Plot $\widehat{f}(x)$ and $M_2 g_2(x)$.

   (c) Compare M_1 and M_2 bounds. Which bound is more convenient for a rejection sampling algorithm? Why?

12. Consider the continuous distribution with density

$$p_1(x) = |\sin x|, \quad -\pi < x < \pi. \tag{1.7}$$

   (a) Determine the target density $p_1(x)$ and the instrumental density $g_1(x)$ for a rejection algorithm. For generating a random variable from the instrumental density, you can use the commands: rnorm (generates a normal distribution), and rbinom (generates a binomial distribution).

   (b) Let M be such that $p_1(x) \leq M g_1(x)$, where $p_1(x)$ and $g_1(x)$ are given (a). Which M is optimal for the rejection sampling?

   (c) Plot $p_1(x)$ and $M g_1(x)$.

2
Monte Carlo Methods

The name "Monte Carlo" sounds glamorous in many ways. According to legend, Stanislaw Ulam (1909-1984), who was in the process of recovering from an illness, was playing solitaire. Pure combinatorial calculations of the success probability of the solitaire game became too lengthy, so he decided to calculate the probability by playing the game repeatedly. He played the game a hundred times and counted the number of successful plays. Monte Carlo established the idea that an integral, in this case the probability of success, can be viewed as an expected value of a function, and it can be estimated statistically.

2.1 Independent Monte Carlo Methods

The aim in using Monte Carlo (MC) methods is to find the value of the integral:

$$\mu = \int f(x)dx < \infty.$$

The general principle of MC:

- Find a suitable factorization for $f(x)$

$$f(x) = h(x)p(x), \qquad (2.1)$$

where $p(x)$ is a density. μ can be interpreted as expected value

$$\mu = Eh(x) = \int h(x)p(x)dx.$$

- Generate an arbitrary length i.i.d. sample from $p(x)$, i.e., $x_1, \ldots, x_n$. The Monte Carlo approximation is the sample estimate of the mean which is given by the average

$$\widehat{\mu} = \frac{1}{n}\sum_{i=1}^{n} h(x_i).$$

Under

$$\sigma^2 = Var(h(x)) = \int (h(x) - \mu)^2 p(x)dx < \infty \qquad (2.2)$$

from the law of large number

$$\widehat{\mu} \to \mu \quad a.s.,$$

and the central limit theorem

$$\frac{\sqrt{n}(\widehat{\mu} - \mu)}{\sigma} \to N(0,1),$$

we have

$$\widehat{\mu} = \mu + \frac{1}{\sqrt{n}}\sigma O_p(1).$$

Note that the rate $\frac{1}{\sqrt{n}}$ cannot be improved for Monte Carlo methods. There are deterministic numerical integration procedures that can reach the rate $O\left(\frac{1}{n^4}\right)$. However, Monte Carlo integration is less expensive for high dimensional functions. The Monte Carlo approximation of the deterministic integral is a random variable, which is close with a high probability to the exact value of the integral. As an example, see Figure 2.1.

2.1 Independent Monte Carlo Methods

Algorithm 2.1 Independent MC

1. Factorize $f(x) = h(x)p(x)$, where p(x) is a density function.
2. Draw $x^{(1)}, \ldots, x^{(n)}$ from distribution $p(x)$.
3. Approximate μ by

$$\widehat{\mu} = \frac{1}{n}(h(x_1) + \ldots + h(x_n)).$$

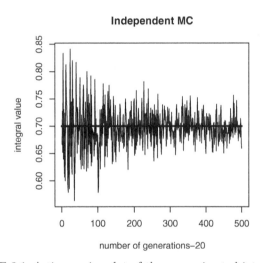

FIGURE 2.1: A time series plot of the approximated integral values.

Example 2.1 (Independent MC) Suppose we want to find the value of

$$\mu = \int_0^{0.4} 4x^5 \exp(-x^3) dx.$$

It holds $\mu = 0.0247$. One can also use MC to calculate μ. In order to calculate the independent Monte Carlo algorithm, we choose a factorization using the Weibull distribution with scale = 1 and shape = 3,

$$p(x) = 3x^2 \exp(-x^3), \text{ for } x \geq 0.$$

Thus

$$h(x) = \frac{4}{3} x^3 I_{[0,0.4]}(x).$$

See Figure 2.6. The R code is given below. □

R Code 2.1.6. Independent MC, Example 2.1

```
meth1<-function(N)
{
 x <- rweibull(N,3,1) ### generates N Weibull random variables
 z <- x[x<0.4] ### truncation
 hz <- z^3*4/3
 return(sum(hz)/N) ### integral value
}
meth1(1000) ### carry out the Monte Carlo approximation
```

For a bounded function f with $0 < f(x) < c, x \in [a,b]$, we can apply the "hit or miss" **Monte Carlo** to approximate

$$\mu = \int_a^b f(x)dx.$$

An example of a hit or miss algorithm is shown in Figure 2.2, for more details see also Bloom (1984) p. 220. We fill the rectangle $R = [0,1] \times [a,b]$ with uniformly distributed independent random variables, and then count the number of points in the area A under the curve. Here, we have $p(x,y) = \frac{1}{c}\frac{1}{b-a}$. Thus, $\mu = (b-a)c \int \int_A p(x,y)dxdy$ with $A = \{(x,y) : y < f(x)\}$.

Algorithm 2.2 "hit or miss" Monte Carlo

1. Generate independently $U_i \sim U_{[0,1]}$, $V_i \sim U_{[0,1]}$ $i = 1,\ldots,n$.
2. Estimate
$$\widehat{\mu} = \frac{\#\{i : cV_i < f(a + U_i(b-a))\}(b-a)c}{n}.$$

There are two aspects that are essential for the factorization in (2.1).

(i) Find a procedure where σ^2 in (2.2) is small (variance reduction methods).

(ii) Find a density where the generation of random variables is easy.

In the next example, two different Monte Carlo methods are compared, and the effect of variance reduction is shown.

2.1 Independent Monte Carlo Methods

Hit or Miss

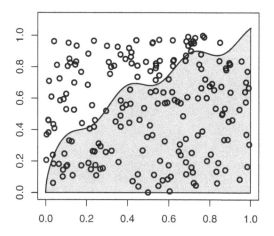

FIGURE 2.2: "Hit or miss" algorithm.

Example 2.2 (Variance reduction) Suppose we want the tail probability

$$\mu = P(C > 2), \quad C \sim Cauchy(0, 1).$$

We propose two independent Monte Carlo methods and compare the variances of the approximation values. Method 1 is a straightforward application of independent MC. We generate independent random numbers $x_1, \ldots, x_n$ from $Cauchy(0, 1)$ and set

$$\widehat{\mu}_{(1)} = \frac{1}{n} \sum_{i=1}^{n} I_{[2,\infty)}(x_i), \text{ where } n\widehat{\mu}_{(1)} \sim Bin(n, \mu),$$

$$Var(\widehat{\mu}_{(1)}) = \frac{1}{n}(\mu(1-\mu)) \approx \frac{1}{n} 0.126.$$

Method 2 explores the symmetry of the Cauchy distribution: $\mu = P(C > 2) = P(C < -2)$, see Figure 2.3. We have

$$\widehat{\mu}_{(2)} = \frac{1}{2n} \sum_{i=1}^{n} I_{(-\infty,-2] \cap [2,\infty)}(x_i), \text{ where } 2n\widehat{\mu}_{(2)} \sim Bin(n, 2\mu),$$

$$Var(\widehat{\mu}_{(2)}) = \frac{1}{4n}(2\mu(1-2\mu)) \approx \frac{1}{n} 0.052.$$

The advantage of Method 2 is illustrated in Figure 2.4. □

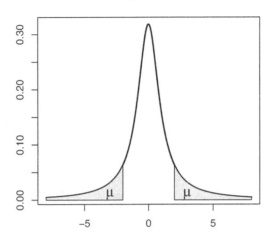

FIGURE 2.3: Symmetry of the Cauchy distribution.

2.1.1 Importance Sampling

The main idea in importance sampling is to focus on the regions of "importance". Importance sampling works as follows: First, we generate a sample from an easy-to-sample trial distribution with density g, we correct the bias by incorporating the importance weights.

For example, see the plot from Example 2.1 and Example 2.3 in Figure 2.6. If we use the Monte Carlo method, only a few points will fall under the right tail of the distribution (Figure 2.6, the right plot), and this area will be underestimated. We can instead use importance sampling and set more weight on the right tail.

Now we present the algorithm. We are interested in the value of $\mu = \int f(x)dx$. First, we factorize $f(x)$ into $h(x)p(x)$. Then:

$$\mu = Eh(x) = \int h(x)p(x)dx.$$

Algorithm 2.3 Importance Sampling

1. Draw $x^{(1)}, \ldots, x^{(m)}$ from a trial distribution g.

2.1 Independent Monte Carlo Methods

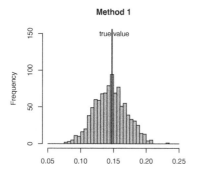
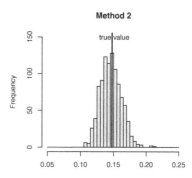

FIGURE 2.4: Variance reduction for Example 2.2. Method 1 relies only on one tail, while Method 2 explores the symmetry of the distribution.

---

2. Calculate the importance weights

$$w^{(j)} = \frac{p(x^{(j)})}{g(x^{(j)})}, \quad j = 1, \ldots, m.$$

3. Approximate μ by

$$\widehat{\mu} = \frac{w^{(1)} h(x^{(1)}) + \ldots + w^{(m)} h(x^{(m)})}{w^{(1)} + \ldots + w^{(m)}}.$$

For more details, see Lui (2001).

Why does importance sampling work?
Applying the law of large numbers to

$$\frac{1}{m} \sum_{j=1}^{m} w^{(j)} h(x^{(j)}) \to E_g w(x) h(x),$$

with

$$w(x) = \frac{p(x)}{g(x)},$$

it holds

$$E_g w(x) h(x) = \int w(x) h(x) g(x) dx = \int \frac{p(x)}{g(x)} h(x) g(x) dx = \int h(x) p(x) dx = \mu.$$

Then, we apply the law of large numbers to

$$\frac{1}{m} \sum_{j=1}^{m} w^{(j)} \to E_g w(x)$$

with
$$E_g w(x) = \int w(x)g(x)dx = \int \frac{p(x)}{g(x)} g(x) dx = \int p(x) dx = 1.$$
Hence
$$\widehat{\mu} = \frac{\frac{1}{m}\sum_{j=1}^m w^{(j)} h(x^{(j)})}{\frac{1}{m}\sum_{j=1}^m w^{(j)}} \to \mu \ \ a.s.$$

Example 2.3 (Importance sampling) Consider the integral in Example 2.1, $\mu = \int_0^{0.4} 4x^3 \exp(-x^3) dx$. We choose $p(x) = 3x^2 \exp(-x^3)$, that is $Weib(a,b)$ with shape $a = 3$ and scale $b = 1$, and $h(x) = \frac{4}{3} x I_{[0,0.4]}(x)$. We take $Exp(\lambda)$ with $\lambda = 1$, $g(x) = \exp(-x)$ as instrumental distribution. Then the weight function is $w(x) = \frac{p(x)}{g(x)} = 3x^2 \exp(-x^3 + x)$. See Figures 2.5 and 2.6 and R code 2.1.7. □

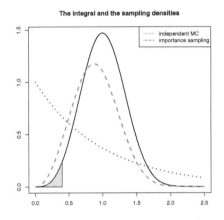

FIGURE 2.5: Example 2.1 and Example 2.3, comparison of Monte Carlo with importance sampling.

2.1.2 The Rule of Thumb for Importance Sampling

We start by quoting Theorem 2.5.1 from Lui (2001), p. 31.

Theorem 2.1 Let $f(z_1, z_2)$ and $g(z_1, z_2)$ be two probability densities, where the support of f is a subset of the support of g. Then,
$$Var_g\left(\frac{f(z_1, z_2)}{g(z_1, z_2)}\right) \geq Var_g\left(\frac{f_1(z_1)}{g_1(z_1)}\right),$$

2.1 Independent Monte Carlo Methods

> where $f_1(z_1)$ and $g_1(z_1)$ are the marginal densities.

Proof: It holds
$$\frac{f_1(z_1)}{g_1(z_1)} = \int \frac{f(z_1,z_2)}{g_1(z_1)} dz_2.$$

Note, $g(z_1, z_2) = g_1(z_1) g_{2|1}(z_2 \mid z_1)$, where $g_{2|1}(z_2 \mid z_1)$ is the conditional density. Then

$$\int \frac{f(z_1,z_2)}{g_1(z_1) g_{2|1}(z_2 \mid z_1)} g_{2|1}(z_2 \mid z_1) dz_2 = \int \frac{f(z_1,z_2)}{g(z_1,z_2)} g_{2|1}(z_2 \mid z_1) dz_2$$
$$= E_g\left(\frac{f(z_1,z_2)}{g(z_1,z_2)} \mid z_1\right).$$

Applying
$$Var(r(z_1,z_2)) = E\left(Var(r(z_1,z_2) \mid z_1)\right) + Var(E(r(z_1,z_2) \mid z_1))$$
$$\geq Var(E(r(z_1,z_2) \mid z_1))$$

to $r(z_1,z_2) = \frac{f(z_1,z_2)}{g(z_1,z_2)}$ delivers the statement. $\square$

Note that Theorem 2.1 is based on the Theorem of Blackwell and the method of Blackwellization. Based on this result, Lui formulated a rule of thumb for Monte Carlo computation:

> One should carry out analytical computation as much as possible.

R Code 2.1.7. Importance sampling, Example 2.3.

```
meth2 <- function(N)
{
 y <- rexp(N) ### trial distribution Exp(1)
 w <- y^2*3*exp(-y^3+y) ### weights
 h <- (y^3*4/3)[y<0.4]
 W <- W[y<0.4]
 return(sum(h*W)/sum(w)) ### integral value
}
meth2(1000) ### carry out the method
```

Comparing different Monte Carlo approximations means comparing random numbers, and can be done by comparing the distributions. In Figure 2.4, we compared two histograms. For continuous distributed approximation values, we recommend the use of a violin plot to compare the estimated densities. The command violinplot in R delivers a comprehensive figure, where the densities are plotted upright and normalized by a standard height. Each method

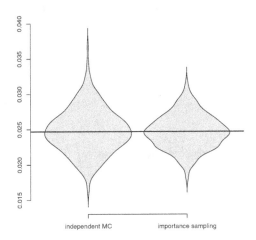

FIGURE 2.6: Comparison of methods by violin plots.

was repeated several times and the density estimation implemented in the violinplot was applied. The importance sampling delivered the best results, where the random approximated values of the integral were more concentrated around the true value.

R Code 2.1.8. Comparison of Monte Carlo methods, Examples 2.1 and 2.3.

```
M1 <- rep(NA,200)
M2 <- rep(NA,200)
for (i in 1:200){
 M1[i] <- meth1(1000)
 M2[i] <- meth2(1000)### run each method 200 times
  }
library(UsingR)
Mcomb <- c(M1,M2)
comb <- c(rep(1,200),rep(2,200))
violinplot(Mcomb~comb,col=grey(0.8),
names=c("independent MC","importance sampling"))
```

2.2 Markov Chain Monte Carlo

Once more we consider the problem of determining the value of the integral $\mu = \int f(x)dx$, where we factorize $f(x) = h(x)\pi(x)$.

$$\mu = E_\pi h(x) = \int_B h(x)\pi(x)dx, \quad B \subseteq \mathbb{R}^d.$$

Markov Chain Monte Carlo (MCMC) methods are especially useful for integrals over high dimensional regions, because simulation of high dimensional i.i.d. random variables can be highly inefficient. It is more convenient to produce a series in a recursive way using the previous members.

The integral μ can be interpreted as the expected value with respect to the stationary distribution π of a Markov chain. By using MCMC, we generate a Markov chain instead of an i.i.d. sample, and instead of the law of large numbers and the central limit theorem, we use the ergodicity properties of Markov chains. In other words, in Monte Carlo, we use the law of large number to show that sum of i.i.d. random variables converges to the mean. For MCMC, we use the ergodic theorem to show that sum of random variables from a Markov chain converges to the mean. We must now generate a Markov chain for a given stationary distribution π.

In a famous paper by Metropolis et al. (1953), an iterative procedure is proposed for generating a Markov chain for a given stationary distribution. The main idea is based on a trial and error procedure. Let us consider at first the illustrative example from Metropolis et al. (1953).

Example 2.4 (Hard shell gas model) The goal is to generate uniform distributed random balls, which do not overlap. The number of balls is fixed.

T Trial step: Move randomly, one random picked ball. The distance to the new position is normally distributed.

E Error step: Test that the ball on the new position does not overlap with the other balls.

□

Try the following R code to get a better understanding of MCMC algorithms. See Figure 2.7, and Figure 2.8.

R Code 2.2.9. Hard shell gas model

```
x <- c(1:4)
y <- c(1:4)
r <- 0.5 ### start: four balls with radius r
s <- sample(1:4)
```

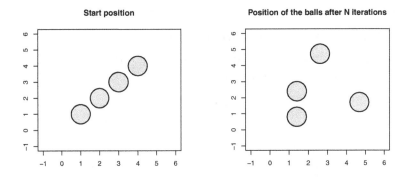

FIGURE 2.7: Hard shell gas model, Example 2.4.

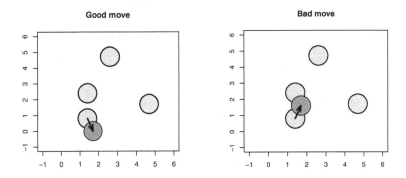

FIGURE 2.8: Accepted new position and rejected position, Example 2.4.

```
s1 <- s[1] ### index of the randomly chosen ball
s2 <- s[2:4] ### indices of the not selected balls
xnew <- x[s1]+2*rnorm(1)
ynew <- y[s1]+2*rnorm(1) ### move the picked ball
## test that the shifted ball do not overlap with the other balls
for (k in s2){
   if((x[k]-xnew)^2+(y[k]-ynew)^2>4*r^2)
   x[s1] <- xnew
   y[s1] <- ynew
 }
### the procedure:
gasmodel <- function(N,x,y,r,a)
{
  n <- length(x)
```

```
for (i in 1:N){
    L <- TRUE
    while(L){
        s <- sample(1:n);s1 <- s[1];s2 <- s[2:n]
        xnew <- x[s1]+a*rnorm(1)
        ynew <- y[s1]+a*rnorm(1)
        for (k in s2){
            if ((x[k]-xnew)^2+(y[k]-ynew)^2>4*r^2)
                {L <- FALSE}}
            x[s1] <- xnew
            y[s1] <- ynew
        }
    }
return(data.frame(x,y))
}
```

Algorithm 2.4 General MCMC

Start with any configuration $x^{(0)}$.

1. Sample a Markov Chain for a given stationary distribution $\pi(x)$:
$$x^{(1)}, ..., x^{(n)},$$
thus
$$\pi(y) = \int A(x,y)\,\pi(x)dx, \qquad (2.3)$$
with
$$A(x,y) = P\left(X^{(1)} = y \mid X^{(0)} = x\right).$$

2. Approximate μ by
$$\widehat{\mu} = \frac{1}{n-m}\sum_{i=m}^{n} h(x^{(i)}).$$

$1,...,m-1$ is the burn-in period.

Why does MCMC work?
For more details, see Lui (2001) pp. 249 and 269.
Suppose the Markov chain is irreducible and aperiodic. Then, it holds for any initial distribution of $X^{(0)}$
$$\widehat{\mu} \to \mu \quad a.s.,$$
and
$$\frac{\sqrt{n}\,(\widehat{\mu} - \mu)}{\sigma_h} \to N(0,1), \quad \text{where } \sigma_h^2 = \sigma^2(1 + 2\sum_{j=1}^{\infty} \rho_j)$$

with
$$\sigma^2 = Var(h(X^{(1)})), \quad \rho_j = Corr(h(X^{(1)}), h(X^{(i+1)})).$$
Thus
$$\widehat{\mu} = \mu + \frac{1}{\sqrt{n}}\sigma_h O_p(1). \tag{2.4}$$

To find a smaller value σ_h, the approximation in (2.4) is better utilized. The problem is now to find a compromise. We want a Markov chain which can be easily generated by using the previous members in the chain but has sufficiently low dependence structure.

A number of the initial iterations of the Markov chain, called a burn-in, should be discarded, see Figure 2.15.

2.2.1 Metropolis-Hastings Algorithm

We must now generate an irreducible, aperiodic Markov chain for a given stationary distribution $\pi(x)$. One approach is to use the Metropolis algorithm, first published in Metropolis et al. (1953). Note that in traditional Markov chain analysis, we are interested in determining $\pi(x)$ for a given $A(x,y)$. Suppose a symmetric proposal distribution $T(x,y) = T(y,x)$ with

$$T(x,y) = P\left(Y = y \mid X^{(t)} = x\right).$$

Algorithm 2.5 Original Metropolis

Start with any configuration $x^{(0)}$.

1. Propose a current state $x^{(t)}$. Sample y from $T(x^{(t)}, y)$.
2. Calculate the ratio
$$R(x^{(t)}, y) = \frac{\pi(y)}{\pi(x^{(t)})}.$$
3. Generate $U \sim U[0,1]$. Update
$$x^{(t+1)} = \begin{cases} y & \text{if } U \leq R(x^{(t)}, y) \\ x^{(t)} & \text{otherwise} \end{cases}.$$

The Metropolis algorithm can be interpreted as "trial-and-error" strategy. The second step is the trial, the third step is the test step see the Figures 2.9 and 2.10.

Hastings generalized this procedure to asymmetric proposal functions in his paper, Hastings (1970). He required only

$$T(x,y) > 0 \text{ iff } T(y,x) > 0.$$

2.2 Markov Chain Monte Carlo

FIGURE 2.9: The trial.

FIGURE 2.10: If the trail does not work, the beetle goes to the old position and starts a new attempt.

Algorithm 2.6 Metropolis-Hastings

Given the current state $x^{(t)}$.

1. Draw y from $T(x^{(t)}, y)$.
2. Calculate the Metropolis-Hastings ratio

$$R(x^{(t)}, y) = \frac{\pi(y)\, T(y, x^{(t)})}{\pi(x^{(t)}) T(x^{(t)}, y)}.$$

> 3. Generate $U \sim U[0,1]$. Update
> $$x^{(t+1)} = \begin{cases} y & \text{if } U \leq \min(1, R(x^{(t)}, y)) \\ x^{(t)} & \text{otherwise} \end{cases}.$$

Why does the Metropolis-Hastings algorithm work?
We have to show that this algorithm produces a Markov chain with $A(x,y)$ such that (2.3) holds. Note that the actual transition function $A(x,y)$ is not equivalent to the trial distribution $T(x,y)$. The new state has to pass the test step. The acceptance probability is $r(x,y) = \min(1, R(x,y))$. Denote the Dirac mass by δ_x, that is $\delta_x(y) = 1$ for $x = y$ and $\delta_x(y) = 0$ otherwise. Furthermore, assume that the trial distribution is continuous, thus $T(x,x) = 0$ and $T(x,y)\delta_x(y) = 0$. We have the actual transition function $A(x,y)$ equal to

$$\begin{aligned}
&= P(\text{``coming from } x \text{ to } y\text{''}) \\
&= P(\text{``coming from } x \text{ to } y\text{''})(1 - \delta_x(y)) \\
&\quad + P(\text{``coming from } x \text{ to } y\text{''})\delta_x(y) \\
&= P(\text{``}y \text{ is proposed and } y \text{ is accepted''})(1 - \delta_x(y)) \\
&\quad + P(\text{``all proposals which are not accepted''})\delta_x(y).
\end{aligned}$$

The step in the algorithm for proposing a new value, and the step for accepting the new value are conditionally independent.

$$\begin{aligned}
P(\text{``}y \text{ is proposed and } y \text{ is accepted''}) &= P(\text{``}y \text{ is proposed''})P(\text{``}y \text{ is accepted''}) \\
&= T(x,y)r(x,y).
\end{aligned}$$

Further

$$\begin{aligned}
P(\text{``all proposals which are not accepted''}) &= \int T(x,y)(1 - r(x,y))dy \\
&= \int T(x,y)dy - \int T(x,y)r(x,y)dy.
\end{aligned}$$

Note $\int T(x,y)dy = 1$ and write

$$r(x) = \int T(x,y)r(x,y)dy.$$

Summarizing, we have

$$A(x,y) = T(x,y)r(x,y)(1 - \delta_x(y)) + (1 - r(x))\delta_x(y).$$

Because $T(x,y)\delta_x(y) = 0$, it holds

$$A(x,y) = T(x,y)r(x,y) + (1 - r(x))\delta_x(y).$$

2.2 Markov Chain Monte Carlo

For more details, see Robert and Casella (1999). $A(x, y)$ is a transition matrix and $\int A(x,y)dx = 1$. Therefore, it is sufficient to show the balance equation

$$\pi(x) A(x, y) = \pi(y) A(y, x), \qquad (2.5)$$

since

$$\int \pi(x) A(x, y) \, dx = \int \pi(y) A(y, x) \, dx = \pi(y) \int A(y, x) \, dx = \pi(y).$$

To show (2.5), we have for the first term of

$$\begin{aligned}
A(x, y) &= \pi(x) T(x, y) r(x, y) \\
&= \pi(x) T(x, y) \min(1, \frac{\pi(y) T(y, x)}{\pi(x) T(x, y)}) \\
&= \min(\pi(x) T(x, y), \pi(y) T(y, x)),
\end{aligned}$$

which is symmetric in x and y. Thus

$$\pi(x) T(x, y) r(x, y) = \pi(y) T(y, x) r(y, x).$$

The second term of $A(x, y)$ is only nonzero for $x = y$, thus, we can exchange them.

$$(1 - r(x))\delta_x(y) = (1 - r(y))\delta_y(x).$$

Hence (2.5) holds.

2.2.2 Special MCMC Algorithms

We introduce different special MCMC algorithms in this section. The key is to find a good trial distribution. In cases of symmetric proposal distributions, one can use a random walk. Assume the target probability $\pi(x)$ is defined on $\mathbb{R}^d$.

Algorithm 2.7 Random-walk Metropolis

Given current state $x^{(t)}$.

    1. Draw $\varepsilon \sim g_\sigma$, where σ is a scaling parameter. Set $y = x^{(t)} + \varepsilon$.

    2. Draw $U \sim U[0, 1]$. Update

$$x^{(t+1)} = \begin{cases} y & \text{if } U \leq \min(1, \frac{\pi(y)}{\pi(x^{(t)})}) \\ x^{(t)} & \text{otherwise} \end{cases}.$$

Example 2.5 (Normal distribution) Given current state $x^{(t)}$.

1. Draw $\varepsilon \sim U(-a, a)$, where $a, a > 0$ is a scaling parameter. Set $y = x^{(t)} + \varepsilon$.
2. Draw $U \sim U[0, 1]$,

$$x^{(t+1)} = \begin{cases} y & \text{if } U \leq \min(1, \exp(-\frac{1}{2}(y^2 - (x^{(t)})^2))) \\ x^{(t)} & \text{otherwise} \end{cases}.$$

Compare Figure 2.11 and Figure 2.12 for different choices of the tuning parameter a. In Figure 2.11, $a = 0.1$ generates a sequence with high dependency and a high acceptance rate. In Figure 2.12, $a = 10$ causes big jumps, a low dependency, and a low acceptance rate. □

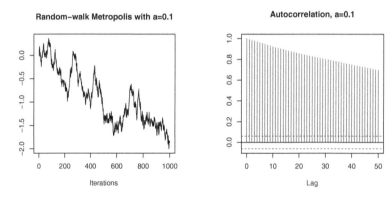

FIGURE 2.11: Random-walk Metropolis, Example 2.5.

R Code 2.2.10. Random-walk Metropolis, Example 2.5

```
MCMC <- function(a,seed,N)
{
    rand <- rep(NA,N)
    rand[1] <- seed
    for(i in 2:N) {
        rand[i] <- seed+a*runif(1,-1,1)
        r <- min(1,exp(0.5*(seed^2-rand[i]^2)))
        if (runif(1)<r){seed <- rand[i]}else{rand[i] <- seed}
        }
  return(rand)
 }
M1 <- MCMC(0.01,0,1000)
```

2.2 Markov Chain Monte Carlo

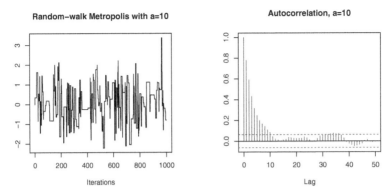

FIGURE 2.12: Random-walk Metropolis, Example 2.5.

```
plot.ts(M1,ylab="",main="Simulation with a=0.01")
acf(M1,lag.max=50) ###  autocorrelation graph
```

The proposal distribution of the next algorithm has no dependency between the members of the chain, but the generated chain does! This algorithm is similar to the rejection algorithm and the importance sampling method; however, the test step and the updating are different. If the new proposal is rejected, the chain stays constant. Also, no bound $\pi(x) < Mg(x)$ is required as for the rejection algorithm.

Algorithm 2.8 Metropolized Independence Sampler (MIS)

Given current state $x^{(t)}$.

1. Draw $y \sim g(y)$.
2. Draw $U \sim U[0,1]$. Update

$$x^{(t+1)} = \begin{cases} y & \text{if } U \leq \min(1, \frac{w(y)}{w(x^{(t)})}) \\ x^{(t)} & \text{otherwise} \end{cases},$$

where

$$w(x) = \frac{\pi(x)}{g(x)}.$$

The following algorithm uses polar coordinates. The direction and the length of the steps are chosen separately, see Figure 2.13. The version given here is also useful for implementing constraints on a sample space. For more details see Givens and Hoeting (2005).

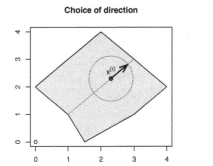 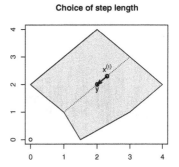

FIGURE 2.13: The hit step and the run step of Algorithm 2.9.

Algorithm 2.9 Hit and Run

Given current state $x^{(t)}$.

1. Draw a random direction $\rho^{(t)} \sim h(\rho)$, where h is a density over the surface of the d-dimensional unit sphere.

2. Find the set of possible step length in the direction $\rho^{(t)}$

$$\Lambda^{(t)} = \left\{\lambda, x^{(t)} + \lambda \rho^{(t)} \in \mathcal{X}\right\}.$$

3. Draw a random length $\lambda^{(t)} \sim g_\lambda^{(t)}(\lambda \mid x^{(t)}, \rho^{(t)})$, where $g_\lambda^{(t)}(\lambda \mid x^{(t)}, \rho^{(t)})$ is a density over $\Lambda^{(t)}$.

4. Propose

$$y = x^{(t)} + \lambda^{(t)} \rho^{(t)}, \text{ such that } y \sim g^{(t)}(y),$$

$g^{(t)}$ is the trial distribution of y given $x^{(t)}$.

5. Calculate the Metropolis-Hastings ratio

$$R(x^{(t)}, y) = \frac{\pi(y) g^{(t)}\left(x^{(t)}\right)}{\pi(x^{(t)}) g^{(t)}(y)}.$$

6. Draw $U \sim U[0,1]$. Update

$$x^{(t+1)} = \begin{cases} y & \text{if } U \leq \min(1, R(x^{(t)}, y)) \\ x^{(t)} & \text{otherwise} \end{cases}.$$

2.2 Markov Chain Monte Carlo

The next algorithm explores the shape of the target distribution. The proposal distribution is a random walk with drift to the mode of the target density, see Figure 2.14. For more details, see Givens and Hoeting (2005), p. 193.

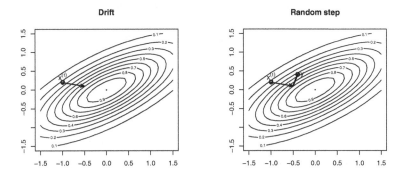

FIGURE 2.14: The drift of the random walk is in direction to the mode of the target distribution, Algorithm 2.10.

Algorithm 2.10 Langevin Metropolis-Hastings

Given current state $x^{(t)}$.

1. Draw $\epsilon^{(t)} \sim N_d(0, I)$ and propose
$$y = x^{(t)} + d^{(t)} + \sigma \epsilon^{(t)},$$
where
$$d^{(t)} = \frac{\sigma^2}{2} \left. \frac{\partial}{\partial x} \ln \pi(x) \right|_{x=x^{(t)}}.$$

2. Calculate the Metropolis-Hastings ratio
$$R(x^{(t)}, y) = \frac{\pi(y) \exp(-\frac{1}{2\sigma^2}(x^{(t)} - y - d^{(t)})^T (x^{(t)} - y - d^{(t)}))}{\pi(x^{(t)}) \exp(-\frac{1}{2\sigma^2}(y - x^{(t)} - d^{(t)})^T (y - x^{(t)} - d^{(t)}))}.$$

3. Draw $U \sim U[0,1]$. Update
$$x^{(t+1)} = \begin{cases} y & \text{if } U \leq \min(1, R(x^{(t)}, y)) \\ x^{(t)} & \text{otherwise} \end{cases}.$$

The approximation of the integral is more "precise" for Markov chains with low dependence between the members. The following advanced algorithm proposes several values in the same step, and makes a suitable choice for one of them. For more details, see Lui (2001). The intention is to produce a Markov chain with low dependence.

Algorithm 2.11 Multiple-try Metropolis-Hastings

Suppose a symmetric weight function $\lambda(x,y)$ and set
$$\omega(x,y) = \pi(y)T(x,y)\lambda(x,y).$$

Given the current state $x^{(t)}$.

1. Sample k i.i.d. proposals from $T(x^{(t)}, y) : y_1, \ldots, y_k$
2. Select one proposal y_{j^*} from $\{y_1, \ldots, y_k\}$ with a probability proportional to $\omega(x^{(t)}, y_{j^*})$.
3. Draw new independent proposals $y_l^* \sim T(y_{j^*}, y), l = 1, \ldots, k-1$, where y_{j^*} stands as the current state. Set $y_k^* = x^{(t)}$.
4. Calculate the generalized Metropolis-Hastings ratio
$$R_g = \frac{\sum_{j=1}^k \omega(x^{(t)}, y_j)}{\sum_{i=1}^k \omega(y_{j^*}, y_i^*)}.$$
5. Draw $U \sim U[0,1]$. Update
$$x^{(t+1)} = \begin{cases} y_{j^*} & \text{if } U \leq \min(1, R_g) \\ x^{(t)} & \text{otherwise} \end{cases}.$$

2.2.3 Adaptive MCMC

The main idea of adaptive MCMC is to adapt the proposal distribution in the Metropolis-Hastings algorithm to the data, which increases the effectiveness of the simulation. This can be done, for instance, by plugging in estimates for unknown free parameters of the proposal distribution. The generated samples have no longer a Markov property. But, it is possible to construct the algorithm in such a way that the ergodic properties remain; for more details, see Atchade and Rosenthal (2005).

Assume the target probability $\pi(x)$ is defined on $\mathbb{R}^d$. In the following algorithm, the empirical covariance matrix

$$Cov(x^{(0)}, \ldots, x^{(t)}) = \frac{1}{t}\sum_{i=0}^{t} x^{(i)} x^{(i)T} - (t+1)\overline{x}\,\overline{x}^T.$$

is used for tuning the proposal distribution.

2.2 Markov Chain Monte Carlo

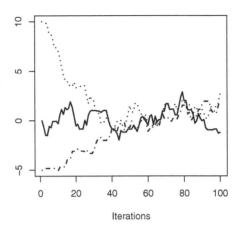

FIGURE 2.15: Different seeds for studying the burn-in time in Example 2.5.

Algorithm 2.12 Adapted Random-walk Metropolis

Given the current states $x^{(0)}, \ldots, x^{(t)}$.

1. Draw $\varepsilon \sim N_d(0, C_{(t)})$, where

$$C_{(t)} = \begin{cases} C_0 & if \quad t \leq t_0 \\ s_d Cov(x^{(0)}, \ldots, x^{(t)}) + s_d \epsilon I_d & otherwise \end{cases},$$

s_d is a scaling parameter, $\epsilon > 0$.
Set $y = x^{(t)} + \varepsilon$.

2. Draw $U \sim U[0,1]$ and update

$$x^{(t+1)} = \begin{cases} y & if \quad U \leq \min(1, \frac{\pi(y)}{\pi(x^{(t)})}) \\ x^{(t)} & otherwise \end{cases}.$$

2.2.4 Perfect Simulation

When running a MCMC sampler, there is always a waiting time for the equilibration - the so-called "burn-in" time. Samples obtained after this period can be considered as samples from the target distribution and can be used for a Monte Carlo estimation. The perfect simulation algorithm was proposed by

Propp and Wilson (1996) and is also called the Propp-Wilson algorithm or "coupling from the past". The main feature of this algorithm is the delivery of the samples from the target distribution without burn-in time. For more details, see Lui (2001) p. 284.

Suppose the target distribution is defined on a finite state space $S = \{s_1, \ldots, s_k\}$.

Algorithm 2.13 Perfect Simulation

1. Set $m = 1$.

2. Draw independently $U_{-N_m+1}, U_{-N_m+2}, \ldots, U_{-1}, U_0$ from $U[0,1]$.

For each $s \in \{s_1, \ldots, s_k\}$ simulate a Markov chain at starting time $-N_m$ in s and running up to 0 by using the $U_{-N_m+1}, U_{-N_m+2}, \ldots, U_{-1}, U_0$ (the same for all chains) in the test step of the Markov chain simulation.

3. If all k chains in Step 2 end up in the same state s' at time 0, then return s' and stop.

4. Increase m by 1 and continue with Step 2.

2.2.5 The Gibbs Sampler

We use Gibbs sampler to sample from a given distribution function if conditional distributions are known. The name Gibbs sampler comes from the Gibbs distribution. Stuart and Geman introduced this procedure in Geman and Geman (1984) for simulating realizations of a Gibbs distribution. The principle of Gibbs sampler is a general one, and is useful for the following setting. Suppose that the random variable can be written as

$$X = (X_1, \ldots, X_p),$$

where the X_is are either one or multidimensional. Moreover, suppose that we can simulate the corresponding conditional distributions $\pi_1, \ldots, \pi_p$; that we can simulate

$$X_i \mid x_{[-i]} \sim \pi_i \left(x_i \mid x_{[-i]} \right), \text{where } x_{[-i]} = (x_1, \ldots, x_{i-1}, x_{i+1}, \ldots, x_p).$$

Algorithm 2.14 Random-Scan Gibbs Sampler

Given the current state $x^{(t)} = (x_i^{(t)})_{i=1,\ldots,p}$.

2.2 Markov Chain Monte Carlo

> 1. Randomly select a coordinate i from $1, \ldots, p$ according to a given probability distribution $p_1, \ldots, p_p$ on $1, \ldots, p$.
> 2. Draw $x_i^{(t+1)}$ from $\pi_i\left(. \mid x_{[-i]}^{(t)}\right)$.
> 3. Set $x_{[-i]}^{(t+1)} = x_{[-i]}^{(t)}$.

Alternatively, one can also change all components one after the other.

> **Algorithm 2.15 Systematic-Scan Gibbs Sampler**
>
> Given the current state $x^{(t)} = (x_i^{(t)})_{i=1,\ldots,p}$. For $i = 1, \ldots, p$ draw $x_i^{(t+1)}$ from
> $$\pi_i\left(. \mid x_1^{(t+1)}, \ldots, x_{i-1}^{(t+1)}, x_{i+1}^{(t)}, \ldots, x_p^{(t)}\right).$$

The densities $\pi_1, \ldots, \pi_p$ are called full conditionals, the only densities used for simulation. Thus, even in high-dimensional problems, all simulations can be univariate! For more details, see Lui (2001) and Robert and Casella (1999).

The systematic-scan Gibbs sampler can be beneficial for bivariate distributions. Here is the illustrative example of the normal distribution.

Example 2.6 (Two-dimensional normal distribution) The aim to find a generator for

$$(X, Y)^T \sim N_2\left(0, \begin{pmatrix} 1 & \rho \\ \rho & 1 \end{pmatrix}\right),$$

where the conditional distribution is

$$X \mid y \sim N(\rho y, 1 - \rho^2).$$

Then, the Gibbs sampler is: For a current state $(x^{(t)}, y^{(t)})^T$

1. Generate

$$X^{(t+1)}/y^{(t)} \sim N\left(\rho y^{(t)}, 1 - \rho^2\right).$$

2. Generate

$$Y^{(t+1)}/x^{(t+1)} \sim N\left(\rho x^{(t+1)}, 1 - \rho^2\right).$$

By iterated calculation, we get

$$\begin{pmatrix} X^{(t)} \\ Y^{(t)} \end{pmatrix} \sim N_2\left(\begin{pmatrix} \rho^{2t-1} x^{(0)} \\ \rho^{2t} y^{(0)} \end{pmatrix}, \begin{pmatrix} 1 - \rho^{4t-2} & \rho - \rho^{4t-1} \\ \rho - \rho^{4t-1} & 1 - \rho^{4t} \end{pmatrix}\right).$$

For $|\rho| < 1$, ρ^{4t} converges to zero for $t \to \infty$. Thus, in a long run, the right distribution is sampled, see Figure 2.16.

□

Exercise 2.1 Prove the formula above (joint distribution of $X^{(t)}$ and $Y^{(t)}$).

R Code 2.2.11. Gibbs sampler for bivariate normal variables, Example 2.6.

```
gibbs-norm<-function(rho,N)
{
 x <- rep(0,N)
 y <- rep(0,N)
  for(i in 1:N){x[i] <- rnorm(1,rho*y[i],(1-rho^2))
      y[i] <- rnorm(1,rho*x[i],(1-rho^2))
      }
return(data.frame(x,y))
}
```

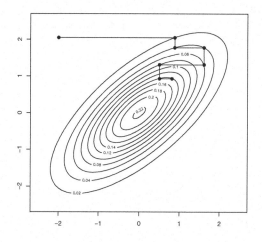

FIGURE 2.16: Generated sample of bivariate normal distribution with $\rho = 0.7$ in Example 2.6.

Why does the Gibbs sampler work?
Let $i = i_t$ be the index chosen at time t. Then, the actual transition matrix is

$$A_t(x,y) = \begin{cases} \pi_i\left(X_i^{(t)} = y_i \mid X_{[-i]}^{(t-1)} = x_{[-i]}\right) & \text{for } x,y, \text{ with } x_{[-i]} = y_{[-i]} \\ 0 & \text{otherwise} \end{cases}.$$

2.2 Markov Chain Monte Carlo

The chain is non-stationary, but aperiodic, and irreducible for $\pi(\omega) > 0$, for all ω. Moreover, $\pi(y) = \sum_x A_t(x,y)\pi(x)$, because

$$\begin{aligned}
\sum_x A_t(x,y)\pi(x) &= \sum_{x_i} \pi_i\left(X_i^{(t)} = y_i \mid X_{[-i]}^{(t-1)} = y_{[-i]}\right)\pi(x) \\
&\quad (\text{Using } x = (y_1, \ldots, y_{i-1}, x_i, y_{i+1}, \ldots, y_n)) \\
&= \pi_i\left(X_i^{(t)} = y_i \mid X_{[-i]}^{(t-1)} = y_{[-i]}\right)\sum_{x_i}\pi_i(x_i \mid y_{[-i]})\pi_i(y_{[-i]}) \\
&= \pi_i\left(X_i^{(t)} = y_i \mid X_{[-i]}^{(t-1)} = y_{[-i]}\right)\pi_i(y_{[-i]})\sum_{x_i}\pi_i(x_i \mid y_{[-i]}) \\
&= \pi_i\left(X_i^{(t)} = y_i \mid X_{[-i]}^{(t-1)} = y_{[-i]}\right)\pi_i(y_{[-i]}) = \pi(y).
\end{aligned}$$

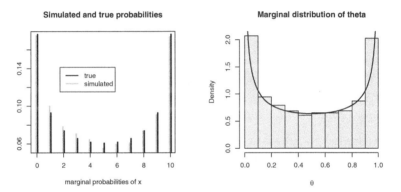

FIGURE 2.17: The values are generated by the Gibbs sampler for Example 2.7.

Gibbs samplers are very useful in Bayesian analysis. For illustration, let us consider the following example.

Example 2.7 (Beta-binomial distribution) Consider an observation x from $Bin(n, \theta)$, and suppose for θ a prior distribution $Beta(\alpha, \beta)$ with density

$$p(\theta) = \frac{1}{beta(a,b)}\theta^{a-1}(1-\theta)^{b-1}.$$

Then, it is known that the posterior distribution is

$$\theta \mid x \sim Beta(\alpha + x, n - x + \beta).$$

Here, we generate the joint distribution of (X, θ) using the conditional distribution of $\theta \mid x$ and of $x \mid \theta$. The marginal distribution of θ is the prior distribution $Beta(\alpha, \beta)$, while the marginal distribution of X is a beta-binomial distribution with probabilities

$$P(X = k) = \binom{n}{k} \frac{beta(k+\alpha, n-k+\beta)}{beta(\alpha, \beta)}.$$

Algorithm: Given the current state $(x^{(t)}, \theta^{(t)})$.

1. Draw $\theta^{(t+1)}$ from $Beta(\alpha + x^{(t)}, n - x^{(t)} + \beta)$.
2. Draw $x^{(t)}$ from $Bin(n, \theta^{(t+1)})$.

In Figure 2.17, 5000 simulated values are compared with the true distributions for $\alpha = \beta = 0.5$. □

R Code 2.2.12. Gibbs sampler for a beta-binomial distribution, Example 2.7.

```
gibbs-beta <- function(a,b,n,N)
{
theta <- rep(0,N)
x <- rep(0,N)
theta[1] <- rbeta(1,a,b)
x[1] <- rbinom(1,n,theta[1])
for(i in 2:N){
    theta[i] <- rbeta(1,a+x[i-1],b+n-x[i-1])
    x[i] <- rbinom(1,n,theta[i])
    }
return(data.frame(x,theta))
}
```

Note that in some cases, MCMC can take an incredibly long time to reach equilibrium. In other words, the chain does not reach equilibrium in practice, and the result will be biased.

2.3 Approximate Bayesian Computation Methods

Approximate Bayesian computation (ABC) is a class of computational methods based on the idea of rejection sampling. Assume a Bayesian model. The data X is generated from the model M with likelihood $p(X|\theta)$ with the parameter θ. The parameter θ comes from a prior distribution $\pi(\theta)$. The goal is to study the posterior of θ:

$$p(\theta|X) = \frac{p(X|\theta)\pi(\theta)}{p(X)}.$$

2.3 Approximate Bayesian Computation Methods

Our primary goal is to generate a sequence of parameters coming from the posterior distribution. The ABC method is beneficial in situations where we can generate data for a given parameter, but we cannot directly work with the likelihood. It can be computationally expensive to calculate the likelihood $p(X|\theta)$, or the likelihood cannot be calculated explicitly (it is intractable). A toy example for this setup is the following example, tracking a fish.

Example 2.8 (Tracking a fish)

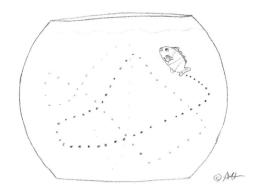

Let us study the movement of a fish in a bowl. Its position $x(t) = (x_1(t), x_2(t), x_3(t))$ is observed at time points $t \in (t_1, \ldots, t_n)$. We assume that
$$x(t) = x(t-1) + R_t \delta_t + \varepsilon_t,$$
where $\delta_t = (\sin(\theta_t)\cos(\phi_t), \sin(\theta_t)\sin(\phi_t), \cos(\theta_t))$ with $\|\delta_t\| = 1$ describe the direction, R_t are the distances, and ε_t are independent normal distributed random variables with variance σ^2. The fish avoids the boundary. The following toy model is proposed:
$$\theta_t \sim 2\pi\, beta(\alpha_1, \beta_1); \phi_t \sim 2\pi\, beta(\alpha_2, \beta_2), R_t \sim U(0, d).$$

When the fish feels that the next step will bring her too near to the border, she forgets her plan and makes a new decision until she can continue her path without danger. The unknown parameter ϑ consists of $\alpha_1, \beta_1, \alpha_2, \beta_2, d$. □

One can use the idea of rejection sampling to generate samples from the posterior of θ, as follows:

1. Generate θ_{new} from $\pi(.)$.
2. Simulate new data X_{new} for the newly generated parameter θ_{new}.
3. Compare the new data X_{new} with the observed data X. If there is "no difference", accept the parameter θ_{new}; otherwise, reject and go back to Step 1.

The accepted θs will have the distribution $p(X|\theta)$.

Algorithm 2.16 Basic ABC Method

Data X is generated from a model M with the parameter θ.

1. Generate θ' from $\pi(.)$.
2. Simulate X' from the model M with the parameter θ'.
3. Accept θ' if $X' = X$.
4. Return to 1.

When X is high dimensional or continuous, the acceptance rate of the Basic ABC method can be very low. In order to measure the difference between data sets, we can use sufficient statistics.

Algorithm 2.17 ABC Method with Sufficient Statistics

Assume $S = (S_1, S_2, \ldots, S_p)$ is sufficient for a model M, and data X is generated from the model M with the parameter θ.

1. Generate θ' from $\pi(.)$.
2. Simulate X' from the model M with the parameter θ' and compute the sufficient statistics S'.
3. Calculate $d(S, S')$, where d is a metric.
4. Accept θ' if $d \leq \epsilon$.
5. Return to 1.

The Euclidean distance can be used as the metric in ABC algorithm. The ABC method can also be used within the MCMC algorithm. The idea is to avoid generating samples from the prior (Step 1 in the ABC algorithm) and instead use MCMC. For more details, see Marjoram et al. (2003).

Algorithm 2.18 ABC-MCMC

Given current parameter θ.

1. Propose θ' from $T(\theta, \theta')$, (for instance $\theta' = \theta + \delta$).
2. Simulate X' from $p(X|\theta')$.
3. Compare X' with X. In case of a good fit, continue; otherwise, go back to Step 1.

2.3 Approximate Bayesian Computation Methods

---

4. Calculate the Metropolis-Hasting ratio

$$R = min(1, \frac{\pi(\theta')T(\theta',\theta)}{\pi(\theta)T(\theta,\theta')}).$$

5. Accept θ' with probability R. Otherwise stay at θ.

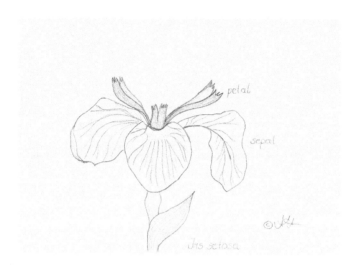

Example 2.9 (ABC method for Iris data) We extract data from the Iris data set. Iris data was collected to classify the morphologic variation of three species of Iris flowers: Setosa, Versicolor, and Virginica. The data contains measurements for the length and the width of the *sepals* and *petals* for each sample. In this example, we use the data for the length and the width of *sepal* of Setosa, a total of 50 samples. We define two variables, V_1 and V_2, one for the length and one for the width. We set a truncated normal distribution for the prior of the means, and uniform distributions for the variances:

$$\begin{aligned}
\mu_1 &\sim Truncated-N(0,\infty,3,1) \\
\mu_2 &\sim Truncated-N(0,\infty,2,1) \\
S_1 &\sim U(0.2,2) \\
S_2 &\sim U(0.2,2) \\
r &\sim U(0,0.9) \\
\Sigma &= \begin{pmatrix} S_1^2 & rS_1S_2 \\ rS_1S_2 & S_2^2 \end{pmatrix}
\end{aligned}$$

Then, the likelihood is a bivariate truncated normal distribution, $Truncated-MVN((\mu_1, \mu_2), \Sigma)$. We use the mean as the sufficient statistics and use the ABC method, Algorithm 2.17, to generate samples for the posterior distribution of the mean, variance, and covariance. For more details see Figure 2.18, and the R code below. □

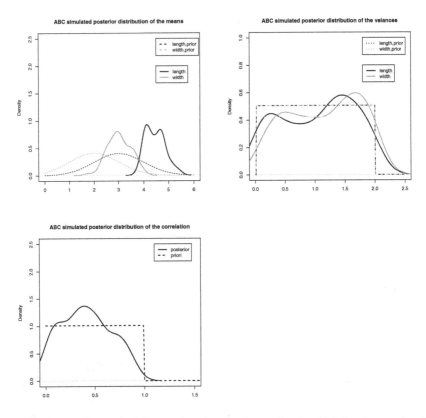

FIGURE 2.18: Example 2.9, simulated posterior and prior distributions using the ABC method.

R Code 2.3.13. Iris data, Example 2.9.

```
library(MASS)
library(tmvtnorm)### library for truncated multivariate normal
library(truncnorm)### library for truncated normal distribution
data(iris)
iris
```

2.3 Approximate Bayesian Computation Methods

```
### subset only sepal measurements of iris.setosa
V1 <- iris[1:50,1] ### length of sepal
V2 <- iris[1:50,2] ### width of sepal
m1 <- mean(V1)
m2 <- mean(V2)

### boundaries for the one sided truncated normal distribution
a <- c(0,0)
### boundaries for the one sided truncated normal distribution
b <- c(Inf,Inf)

### abc-rejection
abc.rej <- function(N,tol){
  rand1 <- rep(NA,N)
  rand2 <- rep(NA,N)
  rand3 <- rep(NA,N)
  rand4 <- rep(NA,N)
  rand5 <- rep(NA,N)

  for(i in 1:N){
    L <- TRUE
    while(L){
      rand1[i] <- rtruncnorm(1,0,Inf,3,1)
      rand2[i] <- rtruncnorm(1,0,Inf,2,1)
      rand3[i] <- runif(1,0.02,2)
      rand4[i] <- runif(1,0.02,2)
      rand5[i] <- runif(1,0,0.9)
      S <- matrix(c(rand3[i]^2,rand5[i]*rand3[i]
      *rand4[i],rand5[i]*rand3[i]*rand4[i],rand4[i]^2),2,2)
      Z <- rtmvnorm(50,c(rand1[i],rand2[i]),S,a,b)
      D <- (mean(Z[,1])-m1)^2+(mean(Z[,2])-m2)^2
      if(D<tol){
        L <- FALSE
      }
    }
  }
  D <- data.frame(rand1,rand2,rand3,rand4,rand5)
  return(D)
}
```

2.4 Problems

1. Calculate
$$\mu = \int_{-2}^{0} \exp(-(x-1)^2) dx$$
with help of the Monte Carlo method:

   (a) Reformulate the integral as an expected value of a normal distribution (use an indicator function).
   (b) Approximate μ by independent MC based on the normal distribution. Generate normally distributed random variables of a suitable size and estimate the related expected value.
   (c) Approximate μ by independent MC based on the uniform distribution.

2. Evaluate the integral of $[\cos(40x) + \sin(10x)]^2$ over $[0, 1]$ using independent Monte Carlo integration methods. Use the command integrate in R to compare your result.

3. Evaluate the integral of $[\sin(x)]^2$ over $[0, \pi]$ using importance sampling. You may use normal distribution as instrumental distribution.

4. Generate a sample from beta distribution $Beta(2, 6)$ using the Metropolis-Hastings algorithm. Use the Kolmogorov-Smirnov test to verify your result.

   (a) Try different starting values and compare the chains by plotting them.
   (b) Plot the autocorrelation functions using the command acf in R.

5. Calculate
$$\int_{-2}^{0} \exp(-(x-1)^2) dx$$
with help of the MCMC method:

   (a) Write the steps of a random walk Metropolis-algorithm for $N(1, 0.5)$.
   (b) What is the proposal distribution $T(x, y)$? Is $T(x, y) = T(y, x)$?
   (c) What is the actual transition function $A(x, y)$ in this example?
   (d) How should the scaling parameter be chosen for maximizing the acceptance probability?
   (e) Generate a Markov chain with stationary distribution $N(1, 0.5)$ by this algorithm.
   (f) Plot this Markov chain as time series (library(ts),ts.plot).

2.4 Problems

(g) Plot the autoregressive correlation function (acf). Discuss the choice of the scaling parameter.

(h) Compute the approximate value of the integral above by MCMC, and compare it with the values from Problem 1.

6. Use the ABC method to simulate a sample from the posterior distribution of the fish movement in Example 2.8.

7. Consider the integral

$$\mu = \int h(x)p(x)dx = \int_0^2 4 - x^2 \sin(x)dx,$$

where $p(x)$ is the density of the uniform distribution over $[0, 2]$.

(a) Apply the importance sampling algorithm with the trial distribution $g(x)$ from $Tri(-3, 3)$. The density of the triangle distribution $Tri(a, b)$ is given by

$$g(x) = \frac{2}{b-a}\left(1 - \frac{2}{b-a}\left|x - \frac{a+b}{2}\right|\right), \quad a < x < b.$$

Write the main steps of the algorithm.

(b) Why does the importance sampling algorithm work?

(c) Consider the rejection algorithm for $p(x)$ based on the trial distribution $g(x)$ of the triangle distribution $Tri(-3, 3)$. Determine M such that $p(x) \leq Mg(x)$.

(d) Compare the importance sampling algorithm in (a) with the independent Monte Carlo method in (c).

8. Consider the integral

$$\mu = \int_{0.3}^1 \int_{0.3}^1 6 \sin(x)^2 \exp(-(y+x))dxdy = k \int_B h(z)\pi(z)dz, \quad (2.6)$$

where $\pi(z) = p(x)p(y)$, with $z = (x, y)$ and $p(x) = exp(-x)$. Apply MCMC for approximating μ.

(a) Write a proposal for $k, B, h(z)$.

(b) Describe a random walk Metropolis algorithm for generating a Markov chain with stationary distribution $\pi(z)$. Write the main steps of the algorithm.

(c) What is the actual transition function in (b)?

(d) What is the acceptance probability in (b)?

(e) How do you approximate the integral in (2.6)?

3
Bootstrap

The bootstrap method was introduced by Efron (1979), "...it is a more primitive method, named the 'bootstrap' for reasons which will become obvious." The use of the term bootstrap derives from the English phrase: "to pull oneself up to by one's bootstrap". The German analogon is described in the "Adventures of Baron Münchhausen". A Chinese student once told me that there is a story of an old Chinese general who was so strong that he could pull himself up in his own armchair. Thus, the idea of bootstrapping is international!

3.1 General Principle

Bootstrap is a general method that can be beneficial in cases where there is no standard procedure to analyze a model. Especially, when the distribution function for the parameter estimator is unknown.

The main principle of the bootstrap is that the data are considered as a good approximation of the population. Therefore, a large number of new observations drawn from this approximated population delivers more than one data set, and gives the possibility for estimating distribution properties. In other words, the bootstrap philosophy is based on the assumption that the generated data have the same properties as the original observed data.

A bootstrap language was invented for describing the method. All inference done for the approximated population is made in the bootstrap world. This means that the bootstrap world is a conditional world given the observed data. The new sample drawn from the approximated population is called the *bootstrap sample*. The generating distribution is the *bootstrap distribution*, which is a conditional distribution given the observed data. The estimate calculated from one bootstrap sample is called the *bootstrap replication*, and is the bootstrapped version of the original estimator. Brown et al. (2001) introduce the word "bootstrappability" (!) for characterizing the ability of the bootstrap method to accurately describe the asymptotic characteristics of the procedure in question. For an overview of the bootstrap notation, see Figure 3.1.

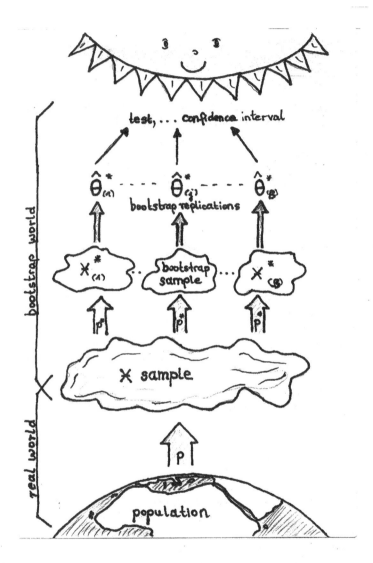

FIGURE 3.1: Bootstrap world. Picture by Annina Heinrich

3.1 General Principle

3.1.1 Unified Bootstrap Framework

Here, we introduce a unified bootstrap framework.

1. **Model**: Suppose $\mathbf{X} = (X_1, \ldots, X_n) \sim \mathbf{P}$ with distribution function $\mathbf{F}$ and $\mathbf{P} \in \{\mathbf{P}_\vartheta, \vartheta \in \Theta\}$, where Θ can be parametric or nonparametric. Note that $\mathbf{X}$ is not necessarily an i.i.d. sample!

2. **Goal**: We are interested in the parameter $\theta = \theta(\mathbf{F})$, the distribution properties of a given estimator $\widehat{\theta} = \widehat{\theta}(\mathbf{X})$, and the distribution properties of a given statistic $T(\mathbf{X}, \mathbf{F})$.

3. **Experiment in the real world**: Getting data $\mathbf{x} = (x_1, \ldots, x_n)$ from $\mathbf{X} = (X_1, \ldots, X_n)$.

4. **Approximation step**: We go over to the bootstrap world. Then, we estimate $\mathbf{F}$ by some "reasonable" estimator $\widehat{\mathbf{F}}_n(\mathbf{z}) = \widehat{\mathbf{F}}(\mathbf{z}, x_1, \ldots, x_n)$, $\mathbf{z} = (z_1, \ldots, z_n) \in \mathbb{R}^n$.

   $\mathbf{P}^*$ is defined as the distribution belonging to $\widehat{\mathbf{F}}_n(\mathbf{z})$, and $\mathbf{P}^*$ we call the *bootstrap distribution*. Note that $\mathbf{P}^*$ is a conditional distribution, given x.

5. **Experiment in the bootstrap world**: We independently draw B random samples $\mathbf{x}^*_{(j)}$ of size n from $\mathbf{P}^*$, $j = 1, \ldots, B$, with

   $$\mathbf{X}^*_{(j)} = (X^*_{1j}, \ldots, X^*_{nj}) \sim \mathbf{P}^* \text{ and } \mathbf{x}^*_{(j)} = (x^*_{1j}, \ldots, x^*_{nj})$$

   $x^*_{(j)}$ are called the *bootstrap samples*.

6. **Estimation in the bootstrap world**: We calculate the estimator for each bootstrap sample $\widehat{\theta}^*_{(j)} = \widehat{\theta}(\mathbf{x}^*_{(j)})$, $j = 1, \ldots, B$. We call them the *bootstrap replications* of $\widehat{\theta}$

   $$\widehat{\theta}^*_{(1)}, \ldots, \widehat{\theta}^*_{(B)}.$$

   The *bootstrap replications* of $T(\mathbf{X}, \mathbf{F})$ are

   $$T^*_{(j)} = T\left(\mathbf{x}^*_{(j)}, \widehat{\mathbf{F}}_n\right), \quad j = 1, \ldots, B.$$

7. **Inference in the bootstrap world**: We treat the bootstrap replications as an i.i.d. sample and apply "usual" statistical procedures such as:

   – Estimating the distribution of $\widehat{\theta}(\mathbf{X})$.
   – Estimating the density $\widehat{\theta}(\mathbf{X})$ using kernel estimate.
   – Estimating quantiles of the distribution of $T(\mathbf{X}, \mathbf{F})$.
   – Calculating intervals for the bias of $\widehat{\theta}(\mathbf{X})$.

- Calculating confidence intervals for different parameters.
- ....

The rule of thumb in bootstrap method: When we consider the statistic $T\left(\widehat{\theta}, \theta\right)$, then the **bootstrapped version** is $T^* = T\left(\widehat{\theta}^*, \widehat{\theta}\right)$. In other words, the estimator $\widehat{\theta}$ is changed to the bootstrapped estimator $\widehat{\theta}^*$ and the true parameter θ is changed to the estimator $\widehat{\theta}$.

Special Cases:
Different versions of bootstrap belong to the unified bootstrap framework. Some of these are:

Efron's bootstrap (simple bootstrap, nonparametric bootstrap, basic bootstrap): In item 1, we assume $\mathbf{X}$ is an i.i.d. sample from $X \sim F$, in item 4, we have

$$\widehat{\mathbf{F}}_n(z) = \prod_{i=1}^{n} \widehat{F}_n(z_i),$$

where $\widehat{F}_n$ is the empirical distribution function (see Efron (1979)). This means the bootstrap sample is drawn with replacement from the observed sample. This is a *resampling method*.

Parametric bootstrap: In item 4, we take $\widehat{\mathbf{F}}_n = \mathbf{F}_{\widehat{\vartheta}}$, where $\widehat{\vartheta}$ is an estimator of a finite dimensional parameter ϑ.

Nonparametric (smoothed) bootstrap: In item 4, we assume $\widehat{\mathbf{F}}_n$ is some nonparametric estimator.

Wild bootstrap: In item 4, the marginal distributions $\widehat{F}_{n,i}$ of $\widehat{\mathbf{F}}_n$ are arbitrary, only the first moments are justified, see Mammen (1993).

Moon bootstrap: (m-out-of-n) In item 5, we assume the bootstrap samples are of size m, $m < n$. There are different sampling strategies possible with and without replacement. Sampling without replacement is also called subsample bootstrap or $\binom{n}{m}$ bootstrap. For more details, see Bickel et al. (1997).

Let us illustrate the bootstrap method step by step using an example from Efron (1979). Reading the original paper of Efron (1979) is useful in understanding the general principle of the boostrap.

Example 3.1 (Sample median)

1. Suppose $\mathbf{X} = (X_1, \ldots, X_n)$ is an i.i.d. sample of size $n = 2k - 1$, from a real valued r.v. X, $X \sim P$ with distribution function F and median $m(F)$.

3.1 General Principle

2. Interested in the bias $T(\mathbf{X},F) = m_{(n)}(\mathbf{X}) - m(F)$ of the sample median $m_{(n)} = m_{(n)}(\mathbf{X}) = X_{[k]}$.

3. Observe $\mathbf{X} = \mathbf{x}$, order the data $x_{[1]} \leq \ldots \leq x_{[i]} \leq \ldots \leq x_{[n]}$.

4. Estimate F by the empirical distribution function $\widehat{F}_n$.

5. For each $j = 1, \ldots, B$ get the bootstrap sample $\mathbf{x}^*_{(j)} = (x^*_{1j}, \ldots, x^*_{nj})$ by resampling x^*_{ij} with replacement from the data $\mathbf{x} = (x_1, \ldots, x_n)$ such that $P^*(x^*_{ij} = x_l) = \frac{1}{n}$. Let N^*_i be the number of copies of x_i in one bootstrap sample $\mathbf{x}^*_j$. Then, the vector

$$(N^*_1, \ldots, N^*_n) \sim multinomial(n, \frac{1}{n}, \ldots, \frac{1}{n}).$$

6. Calculate bootstrap replications

$$T^*_{(j)} = T(\mathbf{x}^*_j, \widehat{F}_n) = x^*_{[k],j} - m_{(n)} \quad j = 1, \ldots, B,$$

where $x^*_{[1],j} \leq \ldots \leq x^*_{[i],j} \leq \ldots \leq x^*_{[n],j}$.

7. Estimate the bias of $m_{(n)}$ by the average of the bootstrap replications $T^*_{(1)}, \ldots, T^*_{(B)}$:

$$\frac{1}{B} \sum_{j=1}^{B} \left(x^*_{[k],j} - m_{(n)} \right).$$

□

Algorithm 3.1 Efron's Bootstrap For Sample Median

Suppose data $\mathbf{x} = (x_1, \ldots, x_n)$ is an i.i.d. sample, where $n = 2k - 1$.

1. For each $j = 1, \ldots, B$, draw a bootstrap sample $\mathbf{x}^*_{(j)} = (x^*_{1j}, \ldots, x^*_{nj})$ by resampling x^*_{ij} with replacement from data $\mathbf{x} = (x_1, \ldots, x_n)$, $i = 1, \ldots, n$.

2. Calculate the median for each bootstrap sample $\widehat{\theta}^*_{(j)} = m_{(n)}(x^*_{(j)}) = x^*_{[k],j}$.

3. Obtain the approximate distribution of the sample median using $\widehat{\theta}^*_{(j)}$'s.

The R code below is to bootstrap sample mean, Example 3.1. Data is simulated from Cauchy distribution. See Figure 3.2.

R Code 3.1.14. Sample median, Example 3.1.

```
x <- rcauchy(11,0,1)### sample from a cauchy distribution
median(x) ### sample median
B<-999
mboot <- rep(1:B)
for(j in 1:B){
    mboot[j] <- median(sample(x,replace=T))
    }
mean(mboot)-median(x) ### bootstrap estimator of the bias
```

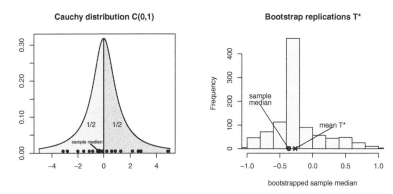

FIGURE 3.2: Example 3.1, bootstrap.

Here is an example with a given data set taken from Efron and Tibshirani (1993).

Example 3.2 (Stroke data) Consider the contingence table:

	success	size
treatment	119	11037
control	98	11034

Let p_1 be the probability of a success (to get a stroke) under the treatment and p_2 the probability of a success in the control group. We are interested in $\theta = \frac{p_1}{p_2}$, especially in testing $H_0 : \theta = 1$. The bootstrap solution for this problem is to derive a bootstrap confidence interval and to check whether 1 is in the interval. Bootstrap sampling is done by sampling from two independent binomial distributions with estimated success probability $\widehat{p}_1 = 119/11037$ and $\widehat{p}_2 = 98/11034$. The bootstrap replications of

$$\widehat{\theta} = \frac{\widehat{p}_1}{\widehat{p}_2}$$

3.1 General Principle

are ordered and the sample quantiles of the bootstrap replications are taken as confidence bounds. For more details, see the following R code and Figure 3.3. Here 1 is included in the interval. Thus, the treatment demonstrates no significant improvement. □

R Code 3.1.15. Stroke data, Example 3.2.

```
p1 <- 119/11037; n1 <- 11037; p2 <- 98/11034; n2 <- 11034
boot1 <- function(p1,n1,p2,n2,B){
    theta <- rep(NA,B);
    for(i in1:B){
        x1 <- rbinom(1,n1,p1)
        x2 <- rbinom(1,n2,p2)
        theta[i] <- (x1*n2)/(x2*n1)
    }
    return(theta)}
R <- sort(boot1(p1,n1,p2,n2,1000))
R[25]; R[975]
```

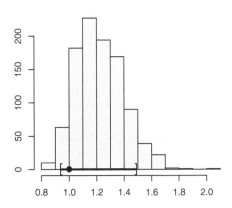

FIGURE 3.3: Example 3.2.

Let us give one more example where the data are not an i.i.d. sample but a time series.

Example 3.3 (Time series)

1. Suppose the data $(y_1, \ldots, y_n) \sim \mathbf{P}_{\alpha,\mu}$ follow an autoregressive time series of first order with

$$y_i = \mu + \alpha y_{i-1} + \varepsilon_i, \quad \varepsilon_i \sim N(0,1) \text{ i.i.d.}, \quad i = 1, \ldots, n.$$

2. We are interested in the distribution properties of

$$\widehat{\alpha} = \frac{\sum_{i=1}^{n-1} (y_i - \overline{y})(y_{i+1} - \overline{y})}{\sum_{i=1}^{n-1} (y_i - \overline{y})^2} \quad \text{where } \overline{y} = \frac{1}{n} \sum_{i=1}^{n} y_i.$$

3. Observe the time series $y_0 = 0, y_1, \ldots, y_n$.
4. Estimate $\mathbf{P}_{\alpha,\mu}$ by $\mathbf{P}_{\widehat{\alpha},\overline{y}}$.
5. For $j = 1, \ldots, B$ generate independently random numbers $\varepsilon_{i,j}^* \sim N(0,1)$, $i = 1, \ldots, n$. Then, the bootstrap samples are iteratively calculated by

$$y_{i,j}^* = \overline{y} + \widehat{\alpha} y_{i-1,j}^* + \varepsilon_{i,j}^*, \quad i = 1, \ldots, n \quad y_0 = 0.$$

6. Calculate the bootstrap replications

$$\widehat{\alpha}_{(j)}^* = \frac{\sum_{i=1}^{n-1} (y_{ij}^* - \overline{y}_j^*)(y_{i+1,j}^* - \overline{y}_j^*)}{\sum_{i=1}^{n-1} (y_{ij}^* - \overline{y}_j^*)^2} \quad \text{where } \overline{y}_j^* = \frac{1}{n} \sum_{i=1}^{n} y_{ij}^*.$$

The simulation of the bootstrap sample in item 5 is called model-based bootstrap, see Figure 3.4. □

The parametric bootstrap approach is not the only way for implementing bootstrap in time series analysis. A blockwise sampling strategy is the best-known method. For more details, see Härdle et al. (2003). In Jentsch and Kreiss (2010), a bootstrap sampling is proposed based on the Fourier transform as well as on the time domain. See Section 3.6 for algorithm regarding time series.

3.1.2 Bootstrap and Monte Carlo

Theoretically, it is possible to directly calculate the distribution of a bootstrapped statistic $T^* = T\left(\mathbf{X}^*, \widehat{\mathbf{F}}_n\right)$. In many cases, the calculation is analytically complicated. However, the distribution of T^* can be approximated by a Monte Carlo simulation. The bootstrap sampling step is repeated B times and the samples $\mathbf{x}_{(1)}^*, \ldots, \mathbf{x}_{(B)}^*$ are generated (item 5 in the general method description), and the bootstrap replications $T_{(1)}^*, \ldots, T_{(B)}^*$ are calculated (item 6 in the general method description). Thus, the general bootstrap method becomes a combination of the bootstrap idea and a Monte Carlo simulation.

3.1 General Principle

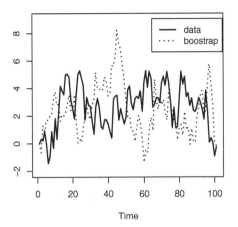

FIGURE 3.4: Example 3.3, time series.

Example 3.4 (Bias of sample mean) This example is taken from Efron (1979). Suppose an i.i.d. sample $(X_1, \ldots, X_n)$ of Bernoulli variables with $\theta = E(X)$. We are interested in the bias

$$T(\mathbf{x}, \mathbf{F}) = \bar{x} - E(X).$$

Draw a bootstrap sample $(X_1^*, \ldots, X_n^*)$ from $Ber(\bar{x})$. Then $\sum_{i=1}^{n} X_i^* \sim P^* = Bin(n, \bar{x})$. The bootstrapped statistic is

$$T^* = T\left(\mathbf{X}^*, \widehat{\mathbf{F}}_n\right) = \frac{1}{n} \sum_{i=1}^{n} X_i^* - \bar{x},$$

and

$$E^*(T^*) = 0, \quad Var^*T^* = \frac{1}{n}\bar{x}(1 - \bar{x}).$$

where E^* is the expected value based on P^*. The mean and the variance of T^* calculated by the Monte Carlo approximation is

$$\frac{1}{B}\sum_{j=1}^{B} T_{(j)}^* = \frac{1}{n}\frac{1}{B}\sum_{j=1}^{B}\sum_{i=1}^{n} x_{ij}^* - \bar{x},$$

with $T^*_{(j)} = T\left(\mathbf{x}^*_{(j)}, \mathbf{F}_n\right)$, and

$$\frac{1}{B}\sum_{j=1}^{B}(T^*_{(j)} - \frac{1}{B}\sum_{j=1}^{B}T^*_{(j)})^2$$
$$= \frac{1}{B}\sum_{j=1}^{B}(\frac{1}{n}\sum_{i=1}^{n}x^*_{ij} - \frac{1}{n}\frac{1}{B}\sum_{j=1}^{B}\sum_{i=1}^{n}x^*_{ij})^2.$$

□

Exercise: Write an R code to calculate the number of bootstrap replications needed for an accurate approximation of $E^*T(\mathbf{X}^*, \widehat{\mathbf{F}}_n)$ and $Var^*T(\mathbf{X}^*, \widehat{\mathbf{F}}_n)$.

Example 3.5 (Sample median) Continuation of Example 3.1. Let us derive the distribution of the bootstrapped bias of the sample median $m_{(n)} = x_{[k]}$, $n = 2k - 1$. Set $B = 1$. Then, the bootstrap replication is

$$m^*_{(n)} = x^*_{[k]} \quad , \quad T^* = T(\mathbf{x}^*, \widehat{F}_n) = x^*_{[k]} - m_{(n)} = x^*_{[k]} - x_{[k]},$$

where $x^*_{[1]} \le \ldots \le x^*_{[i]} \le \ldots \le x^*_{[n]}$. T^* is discrete, distributed on

$$\{x_{[l]} - x_{[k]}, l = 1, \ldots, n\}.$$

Hence we calculate

$$\begin{aligned}P^*\left(T^* = x_{[l]} - x_{[k]}\right) &= P^*\left(T^* > x_{[l-1]} - x_{[k]}\right) - P^*\left(T^* > x_{[l]} - x_{[k]}\right)\\&= P^*\left(x^*_{[k]} > x_{[l-1]}\right) - P^*\left(x^*_{[k]} > x_{[l]}\right).\end{aligned}$$

Note $\#\{j; x^*_j \le x_{[l]}\} \sim Bin(n, \frac{l}{n})$ and $x^*_{[k]} > x_{[l]} \Leftrightarrow \#\{j; x^*_j \le x_{[l]}\} \le k - 1$. Thus

$$P^*\left(T^* = x_{[l]} - x_{[k]}\right) = bin(k-1, n, \frac{l-1}{n}) - bin(k-1, n, \frac{l}{n}).$$

□

As we have shown, it is possible to calculate the distribution of T^* directly, but it can be rather complicated. With replicated bootstrap sampling, we can estimate the distribution of T^* sufficiently precise.

3.1.3 Conditional and Unconditional Distribution

If we draw bootstrap samples from $\mathbf{P}^*$, then the bootstrap replications are independent samples. The question is then "what happens to the 'independent' bootstrap replications, when we go over to the unconditional distributions?".

3.1 General Principle

We calculate the dependence structure of parametric bootstrap samples in a linear normal model.
Consider a linear regression model with a design matrix X of full rank

$$Y = X\beta + \varepsilon, \ \varepsilon \sim N_n(0, \sigma^2 I_n).$$

The variance σ^2 is known. The least squares estimator is $\widehat{\beta} = (X^T X)^{-1} X^T Y$. Note that we use the denotation of linear models, where Y is the column vector of the n observations, and X is the $(n \times p)$ design matrix. Using the parametric bootstrap, the bootstrap samples are

$$Y^*_{(j)} = X\widehat{\beta} + \varepsilon^*_{(j)}, \ \varepsilon^*_{(j)} \sim N_n(0, \sigma^2 I_n), \ j = 1, \ldots, B.$$

Let us derive the unconditional distribution of $\overrightarrow{\mathbf{Y}^*} = (Y_1^{*T}, \ldots, Y_{B^*}^{*T})^T$. Because ε and $\varepsilon^*_{(j)}$ are normally distributed, the distribution of $\overrightarrow{\mathbf{Y}^*}$ is normal too. Thus, it is enough to calculate expectation and covariance matrix. We have

$$E(Y^*_{(j)}) = E\left(E(Y^*_{(j)}|Y)\right) = E(X\widehat{\beta}) = X\beta, \text{ for all } j = 1, \ldots, B.$$

Using the general relation

$$Cov(U, Z) = E_Y(Cov((U, Z)|Y)) + Cov(E(U|Y), E(Z|Y)),$$

we get

$$Cov\left(Y^*_{(k)}, Y^*_{(j)}\right) = E\left(Cov((Y^*_{(k)}, Y^*_{(j)})|Y)\right) + E\left(X\widehat{\beta} - X\beta\right)\left(X\widehat{\beta} - X\beta\right)^T.$$

Because of the conditional independent sampling, we have $Cov(Y^*_{(k)}, Y^*_{(j)}|Y) = 0$ for $j \neq k$. Otherwise, $Cov((Y^*_{(k)}, Y^*_{(k)})|Y) = Cov(\varepsilon^*_{(k)}) = \sigma^2 I_n$. Furthermore,

$$E\left(X\widehat{\beta} - X\beta\right)\left(X\widehat{\beta} - X\beta\right)^T = \sigma^2 X(X^T X)^{-1} X^T.$$

Hence, for $j \neq k$

$$Cov\left(Y^*_{(k)}, Y^*_{(j)}\right) = \sigma^2 X(X^T X)^{-1} X^T,$$

showing that the bootstrap samples are not independent!

Example 3.6 (Illustration of the dependence) We specify the formularies above for the simple linear regression with normal errors and a fixed design

$$y_i = \beta x_i + \varepsilon_i, \ \varepsilon_i \sim N(0, 1) \text{ i.i.d.}, \ i = 1, \ldots, n.$$

Then for $X = (x_1, \ldots, x_n)^T$ and $Y = (y_1, \ldots, y_n)^T$,

$$\widehat{\beta} = d(X) X^T Y \text{ and } \widehat{\beta} \sim N(\beta, d(X)) \text{ with } d(X) = (X^T X)^{-1} = \left(\sum_{i=1}^n x_i^2\right)^{-1}.$$

The first parametric bootstrap sample is generated by
$$Y_{(1)}^* = \widehat{\beta}X + \varepsilon^*, \ \varepsilon^* \sim N_n(0,1), \ Y_{(1)}^* \sim N\left(\beta X, I_n + d(X)XX^T\right).$$
The second bootstrap sample $Y_{(2)}^*$ is generated in the same way and
$$Cov(Y_{(1)}^*, Y_{(2)}^*) = d(X)XX^T.$$
The bootstrap replications $\widehat{\beta}_{(j)}^*$, $j = 1, 2$ are defined by $\widehat{\beta}_{(j)}^* = d(X)X^TY_{(j)}^*$, thus
$$Cov\left(\widehat{\beta}_{(1)}^*, \widehat{\beta}_{(2)}^*\right) = d(X)^2 X^T Cov(Y_{(1)}^*, Y_{(2)}^*)X = d(X)^3 X^T X X^T X = d(X).$$
Furthermore, for $j = 1, 2$
$$Var(\widehat{\beta}_{(j)}^*) = d(X)^2 X^T Cov(Y_{(j)}^*)X = d(X)^2 X^T \left(I + d(X)XX^T\right) X = 2d(X),$$
and hence
$$Corr(\widehat{\beta}_{(1)}^*, \widehat{\beta}_{(2)}^*) = \frac{Cov\left(\widehat{\beta}_{(1)}^*, \widehat{\beta}_{(2)}^*\right)}{\sqrt{Var(\widehat{\beta}_{(1)}^*)}\sqrt{Var(\widehat{\beta}_{(2)}^*)}} = \frac{1}{2},$$
see Figure 3.5. □

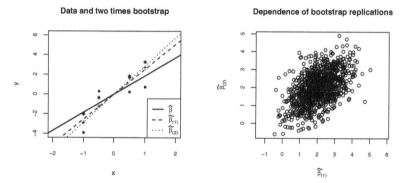

FIGURE 3.5: Simulation study to illustrate Example 3.6. The sample correlation is 0.512.

3.2 Basic Bootstrap

Let $\mathbf{X} = (X_1, \ldots, X_n)$ be an i.i.d. sample with $X \sim P$ and distribution function F. Denote by $\widehat{F}_n$ distribution function of the empirical distribution on

3.2 Basic Bootstrap

$x_1, \ldots, x_n$ putting mass $\frac{1}{n}$ on each x_i. Note that the moments of the empirical distribution are the sample moments. By the law of large numbers, we know that for all x the empirical distribution function $\widehat{F}_n(x)$ converges in probability to $F(x)$, for $n \to \infty$.

3.2.1 Plug-in Principle

The plug-in principle means that by changing to the bootstrap world, we replace F by the empirical distribution function $\widehat{F}_n(x)$. The following table gives a summary of the replacements.

The real world		The bootstrap world
F d.f.	$\rightsquigarrow$	$\widehat{F}_n$ empirical d.f.
$\mathbf{X} = (X_1, \ldots, X_n),\ X \sim F$	$\rightsquigarrow$	$\mathbf{X}^* = (X_1^*, \ldots, X_n^*),\ X^* \sim \widehat{F}_n$
$\theta(F)$	$\rightsquigarrow$	$\theta\left(\widehat{F}_n\right) = \widehat{\theta}(\mathbf{X})$
$T(\mathbf{X}, F)$	$\rightsquigarrow$	$T\left(\mathbf{X}^*, \widehat{F}_n\right)$
$\widehat{\theta}(\mathbf{X})$	$\rightsquigarrow$	$\widehat{\theta}(\mathbf{X}^*) = \widehat{\theta}^*$

For example, consider the r'th moment:

$$\theta(F) = EX^r = \int x^r dF.$$

Then the plug-in estimate is defined as:

$$\widehat{\theta} = \theta\left(\widehat{F}_n\right) = \int x^r d\widehat{F}_n = \frac{1}{n} \sum_{i=1}^{n} x_i^r.$$

Thus all moment estimators are functions of the empirical distribution function and the plug-in principle is applied. The bias of $\widehat{\theta}$ as an estimator of $\theta(F)$ is defined by

$$bias_F = E_F \widehat{\theta} - \theta(F).$$

The bootstrap estimate of the bias is the estimate of

$$bias_{\widehat{F}_n} = E_{\widehat{F}_n} \widehat{\theta}(x^*) - \theta(\widehat{F}_n),$$

given by

$$\widehat{bias}_B = \overline{\theta}^* - \theta(\widehat{F}_n),\quad \overline{\theta}^* = \frac{1}{B}\sum_{j=1}^{B} \widehat{\theta}\left(x_{(j)}^*\right).$$

The accuracy of $\widehat{\theta}$ can be studied by the sample deviation of $\widehat{\theta}\left(x_{(1)}^*\right), \ldots, \widehat{\theta}\left(x_{(B)}^*\right)$:

$$\widehat{se}_B = \left(\frac{1}{B-1} \sum_{j=1}^{B} (\widehat{\theta}\left(x_{(j)}^*\right) - \overline{\theta}^*)^2 \right)^{\frac{1}{2}}.$$

3.2.2 Why is Bootstrap Good?

The following argumentation was first detailed in Singh (1981). Suppose $\mathbf{X} = (X_1, \ldots, X_n)$ is an i.i.d. sample from $X \sim P$, with continuous distribution function F and $0 < Var(X) = \sigma^2 = \sigma^2(F) < \infty$. The parameter of interest is the expectation $\mu = \mu(F) = EX$. The plug-in estimator is the sample mean $\widehat{\mu} = \mu(\widehat{F}_n) = \overline{x}$. Then, by the central limit theorem, it holds

$$Z = \frac{\sqrt{n}\,(\overline{X} - \mu)}{\sigma} \to N(0,1).$$

We rewrite it as $P(Z < z) = \Phi(z) + o(1)$. In the case that the third moment of X exists: $\kappa_3 = \kappa_3(F) = E(X - \mu)^3 < \infty$, we have an Edgeworth expansion:

$$P(Z < z) = \Phi(z) + \frac{1}{\sqrt{n}}\frac{1}{6}\gamma_1(1 - z^2)\varphi(z) + \frac{1}{\sqrt{n}}o(1),$$

where $\varphi(z)$ is the density of $N(0,1)$, and $\gamma_1 = \gamma_1(F) = \frac{\kappa_3}{\sigma^3}$ is the skewness. Applying the plug-in principle, the basic bootstrap replication of

$$Z = \frac{\sqrt{n}\,(\mu(\widehat{F}_n) - \mu)}{\sigma}$$

is

$$Z^* = \frac{\sqrt{n}\,(\overline{x}^* - \mu(\widehat{F}_n))}{\sigma(\widehat{F}_n)}.$$

Analogously, we obtain

$$P^*(Z^* < z) = \Phi(z) + \frac{1}{\sqrt{n}}\frac{1}{6}\widehat{\gamma}_1(1 - z^2)\varphi(z) + \frac{1}{\sqrt{n}}o_P(1).$$

where $\widehat{\gamma}_1 = \gamma_1(\widehat{F}_n)$ is the sample skewness which is asymptotically, and normally distributed. Thus

$$\widehat{\gamma}_1 - \gamma_1 = \sqrt{n}O_p(1).$$

Therefore, we get:

$$\begin{aligned}P(Z < z) - P^*(Z^* < z) &= \Phi(z) - \Phi(z) \\ &+ \frac{1}{6\sqrt{n}}(\widehat{\gamma}_1 - \gamma_1)(1 - z^2)\varphi(z) + \frac{1}{\sqrt{n}}o_P(1) \\ &= \frac{1}{\sqrt{n}}o_P(1).\end{aligned}$$

The conditional distribution $P^*(Z^* < z)$ is therefore a good approximation of the unknown distribution $P(Z < z)$ (asymptotically of the second order when $n \to \infty$). In most cases, $P^*(Z^* < z)$ is not given, but it can be estimated sufficiently good by the bootstrap replications, when $B \to \infty$ (Monte Carlo step).

3.3 Bootstrap Confidence Sets

Reminders:

- x_n is $o_p(1)$ if $x_n \xrightarrow{p} 0$.
- x_n is $O_p(1)$ if $\exists M$, $P(|x_n| < M) \to 0$.

3.2.3 Example where Bootstrap Fails

This example is quoted from Davison and Hinkley (1997), p. 39. Suppose $\mathbf{X} = (X_1, \ldots, X_n)$ is an i.i.d. sample from $U(0, \theta)$. We are interested in θ, where the maximum likelihood estimator of θ is the sample maximum

$$\widehat{\theta}_n = \max(X_i) = X_{[n]}.$$

Note that the limit distribution of $X_{[n]}$ is not normal! Asymptotically, we get an extreme value distribution

$$Q = \frac{n(\theta - X_{[n]})}{\theta} \to Exp(1).$$

Using a basic bootstrap with replication of Q we obtain

$$Q^* = \frac{n(\widehat{\theta}_n - X^*_{[n]})}{\widehat{\theta}_n}.$$

Then

$$\begin{aligned}
P^*(Q^* = 0) &= P^*(X^*_{[n]} = \widehat{\theta}_n) = P^*(\max(X^*_i) = x_{[n]}) \\
&= 1 - P^*(X^*_i < x_{[n]})^n = 1 - (1 - P^*(X^*_i = x_{[n]}))^n \\
&= 1 - (1 - \frac{1}{n})^n,
\end{aligned}$$

and

$$\lim_{n \to \infty} P^*(Q^* = 0) = 1 - \lim_{n \to \infty} (1 - \frac{1}{n})^n = 1 - \frac{1}{e} > 0.$$

This implies that the limit distribution of Q^* cannot be a continuous distribution, see Figure 3.6. This is a contradiction to the bootstrap philosophy that Q^* should have the same limit distribution as Q. For more details, see Belyaev (1995).

3.3 Bootstrap Confidence Sets

The calculation of confidence sets is one of the most important tasks in applied statistics. Nowadays, the bootstrap method is recommended as one of

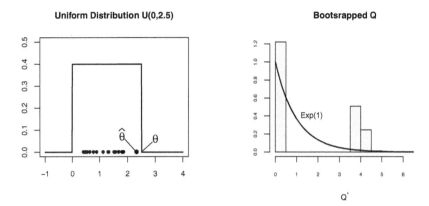

FIGURE 3.6: Example where bootstrap fails.

the universal methods for calculation of confidence sets. Different bootstrap approaches are in this chapter.

Suppose the following setting: $\mathbf{X} = (X_1, \ldots, X_n)$ i.i.d. sample from $X \sim P_\vartheta$. We want a confidence set $C(\mathbf{X})$ for $\theta = g(\vartheta)$, such that for all ϑ

$$P_\vartheta(\theta \in C(\mathbf{X})) = 1 - 2\alpha.$$

3.3.1 The Pivotal Method

The pivotal method delivers a way of using bootstrap for constructing confidence sets. First, let us describe the pivotal method in the "real world". In subsequent subsections, we will present different bootstrap implementations of the pivot principle.

The key is to find a **pivot**. The French word "pivot" is used in many languages to describe an element which is not moving, when everything around it is changing, see Figure 3.7. In statistics, the name pivot is given to a standardized statistic Q depending on $\mathbf{X} = (X_1, \ldots, X_n)$ and $\theta = \theta(F)$ where the distribution of Q is independent of ϑ

$$Q = Q(\mathbf{X}, \theta) \sim P^Q. \tag{3.1}$$

There aren't many situations where it is possible to find a pivot, but here are a few well-known examples.

Example 3.7 (Normal distribution) Assume $\mathbf{X} = (X_1, \ldots, X_n)$ is an i.i.d. sample from $X \sim N(\theta, \sigma^2)$ where σ^2 is known. A pivot is then

$$Q(\mathbf{X}, \theta) = \overline{\mathbf{X}} - \theta \sim N\left(0, \frac{1}{n}\sigma^2\right). \tag{3.2}$$

3.3 Bootstrap Confidence Sets

FIGURE 3.7: The tree in the eye of hurricane as pivot.

In case of an unknown variance, we can estimate the variance by

$$S^2 = \frac{1}{n-1} \sum_{i=1}^{n} (X_i - \overline{\mathbf{X}})^2,$$

and set the student statistic as the pivot,

$$Q(\mathbf{X}, \theta) = \frac{\sqrt{n}\,(\overline{\mathbf{X}} - \theta)}{S} \sim t_{n-1}. \quad (3.3)$$

Assume $\mathbf{X} = (X_1, \ldots, X_n)$ is an i.i.d. sample from $X \sim N(\mu, \sigma^2)$. The parameter of interest is $\theta = \sigma^2$. A pivot statistic is then

$$Q(\mathbf{X}, \theta) = \frac{(n-1)S^2}{\theta} \sim \chi^2_{n-1}. \quad (3.4)$$

□

Example 3.8 (Exponential distribution) Assume $\mathbf{X} = (X_1, \ldots, X_n)$ is an i.i.d. sample from $X \sim Exp(\lambda) = \Gamma(\lambda, 1)$ with $EX = \frac{1}{\lambda}$ and $\lambda X \sim \Gamma(1,1)$. Then, for $\theta = \lambda$

$$Q(\mathbf{X}, \theta) = 2\theta \sum_{i=1}^{n} X_i \sim \chi^2_{2n}. \quad (3.5)$$

□

For more examples see Chapter 2, Section 2.1, in Davison and Hinkley (1997). The general principle for constructing a confidence set using a pivot quantity is as follows. Let the pivot Q be given and let the distribution P^Q be known, such that we have quantiles L and U:

$$P^Q(L \leq Q \leq U) = 1 - 2\alpha.$$

Then, the confidence set is defined by

$$C(\mathbf{X}) = \{\theta : L \leq Q(\mathbf{X}, \theta) \leq U\}.$$

For $\theta \in \mathbb{R}$ and under the assumption of monotony, we can often find the confidence interval $[l(\mathbf{X}), u(\mathbf{X})]$ by

$$L \leq Q(\mathbf{X}, \theta) \leq U \Leftrightarrow l(\mathbf{X}) \leq \theta \leq u(\mathbf{X}).$$

Fortunately, this is the case in the examples above.

Example 3.9 (Normal distribution) Continuation of Example 3.7. Under (3.2) we get the interval

$$\left[\overline{\mathbf{X}} - \frac{\sigma}{\sqrt{n}} z_{1-\alpha},\ \overline{\mathbf{X}} + \frac{\sigma}{\sqrt{n}} z_{1-\alpha} \right],$$

where $z_{1-\alpha}$ is the quantile of the standard normal distribution, $\Phi(z_{1-\alpha}) = 1 - \alpha$.

Under (3.3) we get the student confidence interval

$$\left[\overline{\mathbf{X}} - \frac{S}{\sqrt{n}} t_{n-1,1-\alpha},\ \overline{\mathbf{X}} + \frac{S}{\sqrt{n}} t_{n-1,1-\alpha} \right],$$

where $t_{n-1,1-\alpha}$ is the quantile of the t distribution with $n-1$ degrees of freedom, $P(t \leq t_{n-1,1-\alpha}) = 1 - \alpha$. □

Example 3.10 (Exponential distribution) Continuation of Example 3.8. Under (3.5) we get the interval

$$\left[\frac{\chi^2_{n-1,\alpha}}{2n\overline{\mathbf{X}}},\ \frac{\chi^2_{n-1,1-\alpha}}{2n\overline{\mathbf{X}}} \right],$$

where $\chi^2_{n-1,1-\alpha}$ is the quantile of the chi-squared distribution with $n-1$ degrees of freedom, $P(\chi \leq \chi^2_{n-1,1-\alpha}) = 1 - \alpha$. □

3.3.2 Bootstrap Pivotal Methods

The main complication in using the pivotal method is to find a pivot $Q(\mathbf{X}, \theta)$ with a known distribution P^Q, which is independent of the parameter ϑ. One can use the bootstrap pivotal methods when these two conditions are not fulfilled. We estimate the quantiles of the distribution of Q by bootstrapping. The probability is P^Q_ϑ approximated by the bootstrap distribution P^*,

$$P^Q_\vartheta \left(L^* \leq Q(\mathbf{X}, \theta) \leq U^* \right) \approx P^* \left(L^* \leq Q\left(\mathbf{X}^*, \widehat{\theta}\right) \leq U^* \right) = 1 - 2\alpha,$$

and L^*, U^* are estimated by the percentile method from the bootstrap replications $Q^*_{(j)} = Q\left(\mathbf{X}^*_{(j)}, \widehat{\theta}\right)$. Thus, the bootstrap confidence set is

$$C^*(\mathbf{X}) = \left\{ \theta : Q^*_{[(1+B)\alpha]} \leq Q(\mathbf{X}, \theta) \leq Q^*_{[(1+B)(1-\alpha)]} \right\}.$$

3.3.2.1 Percentile Bootstrap Confidence Interval

A straightforward application of bootstrap gives the following percentile bootstrap confidence interval, see Example 3.2 and Figure 3.3. We suppose $\theta \in \mathbb{R}$. Let $\widehat{\theta}^*_{(j)}$, $j = 1, \ldots, B$ be the bootstrap replications of $\widehat{\theta}$. Let $\widehat{\theta}^*_{[1]} \leq \widehat{\theta}^*_{[2]} \leq \ldots \leq \widehat{\theta}^*_{[j]} \leq \ldots \leq \widehat{\theta}^*_{[B]}$ be the ordered bootstrap replications. Then the percentile bootstrap interval is given by

$$\left[\widehat{\theta}^*_{[(1+B)(\alpha)]}, \widehat{\theta}^*_{[(1+B)(1-\alpha)]} \right]. \tag{3.6}$$

Remark: The estimator $\theta(X, \theta) = \widehat{\theta}$ is used as "pivot".

3.3.2.2 Basic Bootstrap Confidence Interval

We consider $Q(\mathbf{X}, \theta) = \widehat{\theta}_{(j)} - \theta$ to be the "pivot". The bootstrap replications are then $Q^*_{(j)} = \widehat{\theta}^*_{(j)} - \widehat{\theta}$, $j = 1, \ldots, B$ and the basic bootstrap interval is given by

$$\left[2\widehat{\theta} - \widehat{\theta}^*_{[(1+B)(1-\alpha)]}, 2\widehat{\theta} - \widehat{\theta}^*_{[(1+B)\alpha]} \right].$$

3.3.2.3 Studentized Bootstrap Confidence Interval

Also in the general case of an i.i.d. sample of an arbitrary distribution, we take the test statistic of the student test as the "pivot":

$$Z = \frac{\widehat{\theta} - \theta}{\sqrt{\widehat{V}}} = Z(\mathbf{X}, F),$$

where $\widehat{V} = \widehat{V}(\mathbf{X})$ is an estimator of $V = Var(\widehat{\theta})$. We can now draw bootstrap samples $\mathbf{X}^*_{(j)}$ $j = 1, \ldots, B$ and calculate the bootstrap replications for $\widehat{\theta}$ and $\widehat{V}$: $\widehat{\theta}^*_{(j)}, \widehat{V}^*_{(j)}$ $j = 1, \ldots, B$ and set

$$Z^*_{(j)} = \frac{\widehat{\theta}^*_{(j)} - \widehat{\theta}}{\sqrt{\widehat{V}^*_{(j)}}} = Z\left(\mathbf{X}^*_{(j)}, \widehat{F}\right).$$

Let $Z^*_{[1]} \leq Z^*_{[2]} \leq \ldots \leq Z^*_{[B]}$. Then, the studentized bootstrap confidence interval is given by

$$\left[\widehat{\theta} - \sqrt{\widehat{V}} \, Z^*_{[(1+B)(1-\alpha)]}, \widehat{\theta} - \sqrt{\widehat{V}} \, Z^*_{[(1+B)\alpha]} \right].$$

Note, the center and scaling of the interval come from the studentized statistic Z, bootstrap is only used for estimating the quantiles! The studentized bootstrap interval is one of the main used methods today.

The problem is to find good formularies for $\widehat{V}$, an estimator of $V = Var\left(\widehat{\theta}\right)$. One way is to apply the nonparametric delta method, which we will explain shortly. The goal is to find an approximation of the variance of $t(\widehat{F}_n)$, where $\widehat{F}_n$ is the empirical distribution function on $(x_1, \ldots, x_n)$. For $t(G)$, it holds the nonparametric Taylor expansion (for more details, see Davison and Hinkley (1997), p. 45),

$$t(G) \approx t(F) + \int L_t(y, F) dG, \qquad (3.7)$$

where $L_t(y, F)$ is the influence function defined by

$$L_t(y, F) = \lim_{\varepsilon \to 0} \frac{t((1-\varepsilon)F + \varepsilon H_y) - t(F)}{\varepsilon}.$$

H_y is the heaviside step function, given by

$$H_y(u) = \begin{cases} 0 & \text{for } u < y \\ 1 & \text{for } u \geq y \end{cases}.$$

It holds $\int L_t(y, F) dF = 0$. Applying (3.7) on the empirical distribution function $\widehat{F}_n$ we obtain

$$t(\widehat{F}_n) \approx t(F) + \int L_t(y, F) d\widehat{F}_n = t(F) + \frac{1}{n} \sum_{i=1}^{n} L_t(x_i, F).$$

The central limit theorem gives us the approximation

$$t(\widehat{F}_n) \approx t(F) + \frac{1}{\sqrt{n}} \sqrt{V_L(F)} Z, \text{ with } Z \sim N(0, 1),$$

and we can approximate

$$V_L(F) = Var_F(L_t(y, F)) \approx \frac{1}{n} \sum_{i=1}^{n} L_t(x_i, F)^2.$$

Thus for $\widehat{V}(\mathbf{X}) = V(\widehat{F}_n)$ we get

$$\widehat{V}(\mathbf{X}) \approx \frac{1}{n} \sum_{i=1}^{n} L_t(x_i, \widehat{F}_n)^2.$$

The application of this approach is demonstrated in the following "bigcity" example, taken from Davison and Hinkley (1997).

Example 3.11 ("bigcity") The data set consists of paired samples $(u_i, x_i), i = 1, \ldots, n$ i.i.d. of (u, x) with distribution F and expectation (μ_u, μ_x). The $(u_i, x_i), i = 1, \ldots, n$ are the populations of n randomly chosen big cities in the years 1920 and 1930. The parameter of interest is the quotient of the expected values

$$\theta = \frac{E(x)}{E(u)} = \theta(F) = \frac{\int x dF}{\int u dF}.$$

3.3 Bootstrap Confidence Sets

We use the plug-in the estimate of θ

$$\widehat{\theta} = \theta(\widehat{F}_n) = \frac{\int x d\widehat{F}_n}{\int u d\widehat{F}_n} = \frac{\overline{x}}{\overline{u}},$$

where $\widehat{F}_n$ is the empirical distribution function of

$$\begin{array}{cccccc} (u_1, x_1) & \ldots & (u_j, x_j) & \ldots & (u_n, x_n) \\ \frac{1}{n} & \ldots & \frac{1}{n} & \ldots & \frac{1}{n} \end{array}.$$

We apply the nonparametric delta method to

$$t(\widehat{F}_n) = \frac{\int x d\widehat{F}_n}{\int u d\widehat{F}_n} = \frac{\overline{x}}{\overline{u}}.$$

Define H_j as the distribution function of

$$\begin{array}{cccccc} (u_1, x_1) & \ldots & (u_j, x_j) & \ldots & (u_n, x_n), \\ 0 & \ldots & 1 & \ldots & 0 \end{array}$$

and consider

$$t((1-\varepsilon)\widehat{F}_n + \varepsilon H_j) = \frac{(1-\varepsilon)\overline{x} + \varepsilon x_j}{(1-\varepsilon)\overline{u} + \varepsilon u_j} = h_j(\varepsilon).$$

Then the influence function in $y_j = (u_j, x_j)$ is given by

$$L_t(y_j, \widehat{F}_n) = \frac{\partial}{\partial \varepsilon}(h_j(\varepsilon))_{\varepsilon=0} = \frac{x_j - \frac{\overline{x}}{\overline{u}} u_j}{\overline{u}}.$$

Hence the variance of $\frac{\overline{x}}{\overline{u}}$ can be approximated by

$$V_L(\widehat{F}_n) \approx \frac{1}{n} \sum_{i=1}^{n} \left(\frac{x_j - \frac{\overline{x}}{\overline{u}} u_j}{\overline{u}} \right)^2.$$

□

For more details see the R code below and Figure 3.8.

R Code 3.3.16. Confidence intervals for "bigcity", Example 3.11.

```
library(boot)
attach(bigcity)
F <- function(x,y){
    n <- length(x)
    u <- sample(1:n,replace=T)
    data <- data.frame(x[u],y[u])
```

```
        return(data)}
### percentile confidence interval ###
boot <- function(B){
    theta <- rep(NA,B)
    for (i in 1:B){
        B <- F(u,x)
        theta[i] <- mean(B$y)/mean(B$x)}
    return(theta)}
B1 <- sort(boot(999))
LOW1 <- B1[25]
UP1 <- B1[975]
###  basic bootstrap confidence interval ###
theta.hat <- mean(x)/mean(u); B2 <- sort(boot(999))
LOW2 <- 2*theta.hat-B2[975]; UP1 <- 2*theta.hat-B2[25]
### studentized bootstrap confidence interval ###
studb <- function(xb,ub,x,u){
    tb <- mean(xb)/mean(ub)
    varb <- mean((xb-ub*tb)^2)/mean(ub)^2
    studb <- (mean(xb)/mean(ub)-mean(x)/mean(u))/sqrt(varb)
    return(studb)}
bootstud <- function(B){
    TB <- rep(NA,B)
    for (i in 1:B){
        B <- F(u,x)
        TB[i] <- studb(B$y,B$x,x,u)}
    return(TB)}
TS <- sort(bootstud(999));U <- TS[25]; L <- TS[975]
se <- sqrt(mean((x-u*theta.hat)^2)/mean(u)^2)
LOW3 <- theta.hat-se*L; UP3 <- theta.hat-se*U
```

3.3.3 Transformed Bootstrap Confidence Intervals

Let h be a monotone smooth transformation of the parameter of interest θ to $\eta = h(\theta)$. We derive the basic bootstrap confidence interval for $\eta = h(\theta)$ and transform back with the inverse function h^{-1}. Let $h_{[1]}^* = h(\widehat{\theta}_{[1]}^*) \leq \ldots \leq h(\widehat{\theta}_{[B]}^*) = h_{[B]}^*$. Then, the transformed basic bootstrap confidence interval is given by

$$\left[h^{-1}(2h(\widehat{\theta}) - h_{[(1+B)(1-\alpha)]}^*), h^{-1}(2h(\widehat{\theta}) - h_{[(1+B)\alpha]}^*) \right].$$

The same method can also be applied to the studentized bootstrap confidence interval. The studentization comes from the normalization in the central limit theorem. Thus, we are looking for a monotone smooth transformation h of the parameter of interest θ to $\eta = h(\theta)$, such that the distribution of $\frac{\widehat{\eta}-\eta}{\sqrt{V(\eta)}}$

3.3 Bootstrap Confidence Sets

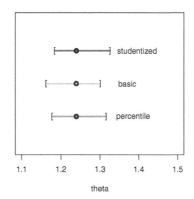

FIGURE 3.8: Example "bigcity".

can be well approximated by a normal distribution. We apply as the "pivot"

$$Z = \frac{h(\widehat{\theta}) - h(\theta)}{\left|h'\left(\widehat{\theta}\right)\right|\sqrt{\widehat{V}(\theta)}} = Z(\mathbf{X}, F),$$

where $\widehat{V}$ is an estimator of $V = Var\left(\widehat{\theta}\right)$. Note that by the delta method $h'\left(\widehat{\theta}\right)^2 \widehat{V}(\theta)$ is an estimator for $V(\eta)$. From bootstrap samples $\mathbf{x}_{(1)}, \ldots, \mathbf{x}_{(B)}$ we calculate $\widehat{\theta}^*_{(j)}, \widehat{V}^*_{(j)}$, $j = 1, \ldots B$. Then, the bootstrap replications of Z

$$Z^*_{(j)} = \frac{h(\widehat{\theta}^*_{(j)}) - h(\widehat{\theta})}{\left|h'\left(\widehat{\theta}^*_{(j)}\right)\right|\sqrt{\widehat{V}^*_{(j)}}} = Z\left(\mathbf{x}^*_{(j)}, \widehat{F}\right)$$

are used for estimating the quantiles: Let $Z^*_{[1]} \leq Z^*_{[2]} \leq \ldots \leq Z^*_{[B]}$. Then the transformed studentized bootstrap confidence interval for θ is given by

$$\left[h^{-1}(h(\widehat{\theta}) - |h'(\widehat{\theta})|\sqrt{\widehat{V}}\, Z^*_{[(1+B)(1-\alpha)]}), h^{-1}(h(\widehat{\theta}) - |h'(\widehat{\theta})|\sqrt{\widehat{V}}\, Z^*_{[(1+B)\alpha]})\right].$$

3.3.4 Prepivoting Confidence Set

This method was first introduced by Beran (1987). It is a consequent application of the pivotal method to bootstrap. The main idea is to generate a pivotal quantity iteratively with an approximative distribution independent of the parameter using Lemma 1.1. Let the statistic $Q(\mathbf{X}, F)$ be a good candidate with similar properties of the "pivot".

Algorithm 3.2 Prepivoting

1. Generate bootstrap replications $Q_{(j)}^* = Q(\mathbf{x}_{(j)}^*, \widehat{F}_n)$, $j = 1, \ldots, B$.
2. Estimate the distribution function F_Q of $Q = Q(\mathbf{X}, F)$ from the bootstrap replications above by $\widehat{F}_Q^*$.
3. Introduce a new pivot
$$R^*(\mathbf{X}, \theta) = \widehat{F}_Q^*(Q(\mathbf{X}, F)),$$
which should be nearly uniform distributed, according to Lemma 1.1.
4. Carry out double bootstrap: Generate bootstrap samples $\mathbf{x}_{(j,k)}^{**}$ from $\mathbf{x}_{(j)}^*$ and calculate replications $R_{(j,k)}^{**} = R_{(j)}^*(\mathbf{x}_{(j)}^*, \theta_{(k)}^*)$ of $R^*(\mathbf{X}, \theta)$, $j = 1, \ldots, B, k = 1, \ldots, B$.
5. Estimate the distribution of R^* from the double bootstrap replications $R_{(j,k)}^{**}$ by $\widehat{P}_{R^*}^*$.
6. Determine an upper U^{**} and a lower bound L^{**} such that
$$\widehat{P}_{R^*}^*(L^{**} \leq R^* \leq U^{**}) \leq 1 - 2\alpha.$$
7. Then the confidence set of $\theta(F)$ is given by
$$\left\{\theta : \widehat{F}_{Q^*}^-(L^{**}) \leq Q(\mathbf{X}, F) \leq \widehat{F}_{Q^*}^-(U^{**})\right\}.$$

This is an iterative method which is computer intensive and still under development.

3.3.5 BC$_a$-Confidence Interval

BC$_a$ stands for bias-corrected and accelerated. The aim is to increase the asymptotic order of the coverage probability such that

$$P_F(\theta \in C^*(\mathbf{X})) = 1 - 2\alpha + O(\frac{1}{n}).$$

This interval was first introduced in Efron (1987), see also Davison and Hinkley (1997). The main idea is to correct the α in the percentile bootstrap interval (3.6). We suppose that there exists a monotone transformation $\eta = h(\theta)$, such that the estimator $\widehat{\eta} = h(\widehat{\theta})$ has a bias ω and acceleration a in the asymptotic

3.3 Bootstrap Confidence Sets

normal distribution, that is

$$\frac{\hat{\eta} - \eta}{\sqrt{V(\eta)}} + \omega \approx N(0,1), \ V(\eta) = (1 + a\eta)^2.$$

Furthermore, we assume that the remainder term is of order n^{-1},

$$P(\frac{\hat{\eta} - \eta}{1 + a\eta} + \omega < z) = \Phi(z) + O(\frac{1}{n}).$$

Hence, we can derive a confidence interval C_η for η

$$P(\eta \in C_\eta) = 1 - 2\alpha + O(\frac{1}{n}),$$

where

$$C_\eta = [\hat{\eta}_{(\alpha)}, \hat{\eta}_{(1-\alpha)}],$$

with

$$\hat{\eta}_{(\alpha)} = \hat{\eta} + (1 + a\hat{\eta})\frac{z_\alpha + \omega}{1 - a(\omega + z_\alpha)},$$

and $\Phi(z_\alpha) = \alpha$. Denote $h^{-1}(\hat{\eta}_{(\alpha)})$ by $\hat{\theta}_{(\alpha)}$. Now, we consider one bootstrap replication $\hat{\eta}^* = h(\hat{\theta}^*)$ and explore the approximation of P by P^*. Because of the monotony of h and using the asymptotic normality of $\hat{\eta}$ we get

$$P^*(\hat{\theta}^* < \hat{\theta}_{(\alpha)}) = P^*(\hat{\eta}^* < \hat{\eta}_{(\alpha)})$$

$$= P^*(\frac{\hat{\eta}^* - \hat{\eta}}{1 + a\hat{\eta}} + \omega < \omega + \frac{z_\alpha + \omega}{1 + a(\omega + z_\alpha)})$$

$$= \Phi(\omega + \frac{z_\alpha + \omega}{1 + a(\omega + z_\alpha)}) + O_{P^*}(\frac{1}{n}).$$

Define the corrected α levels by

$$\alpha_1 = \Phi(\omega + \frac{z_\alpha + \omega}{1 + a(\omega + z_\alpha)}) \quad \alpha_2 = \Phi(\omega + \frac{z_{1-\alpha} + \omega}{1 + a(\omega + z_{1-\alpha})}).$$

Note that the knowledge of the transformation h is not needed, only its existence is required. We apply a bootstrap percentile interval with the corrected α and obtain

$$P(\theta \in \left[\hat{\theta}^*_{[(1+B)\alpha_1]}, \hat{\theta}^*_{[(1+B)\alpha_2]}\right]) = 1 - 2\alpha + O(\frac{1}{n}).$$

It remains to estimate the bias ω and the acceleration a. Consider the likelihood function $l(\eta)$ of $N(\eta - \omega(1 + a\eta), (1 + a\eta)^2)$. The skewness of the score function l' is then

$$\frac{El'(\eta)^3}{(El'(\eta)^2)^{\frac{3}{2}}} = \frac{8a^3 + 6a(1 - \omega a)^2}{(2a^2 + (1 - \omega a)^2)^{\frac{3}{2}}}.$$

We assume that
$$a = O(\frac{1}{\sqrt{n}}), \quad \omega = O(\frac{1}{\sqrt{n}}),$$
and ignore all terms of order $O(\frac{1}{n})$, then the acceleration a is approximated by
$$a \simeq A(F, \eta) = \frac{1}{6} \frac{El'(\theta)^3}{(El'(\theta)^2)^{\frac{3}{2}}}.$$
The skewness is invariant with respect to parametrization, such that $A(F, \eta) = A(F, \theta)$. The acceleration can be estimated by $\tilde{a} = A(\widehat{F}_n, \widehat{\theta})$. Furthermore, because
$$P(\widehat{\theta} < \theta) = P(\widehat{\eta} < \eta) \approx \Phi(\omega),$$
the bias ω can be estimated by bootstrap. We generate bootstrap replications of $\widehat{\theta}$: $\widehat{\theta}^*_{(1)}, \ldots, \widehat{\theta}^*_{(B)}$ and
$$\widetilde{\omega} = \Phi^{-1}(\frac{\#(j : \widehat{\theta}^*_{(j)} < \widehat{\theta})}{B}).$$
Then the corrected levels are calculated by
$$\widetilde{\alpha}_1 = \Phi\left(\widetilde{\omega} + \frac{z_\alpha + \widetilde{\omega}}{1 - \widetilde{a}(z_\alpha + \widetilde{\omega})}\right), \quad \widetilde{\alpha}_2 = \Phi\left(\widetilde{\omega} + \frac{z_{1-\alpha} + \widetilde{\omega}}{1 - \widetilde{a}(z_{1-\alpha} + \widetilde{\omega})}\right).$$
The BC_a-interval is given by
$$\left[\widehat{\theta}^*_{[(1+B)\widetilde{\alpha}_1]}, \widehat{\theta}^*_{[(1+B)\widetilde{\alpha}_2]}\right].$$

In DiCicco and Efron (1996) they introduced an analytical version of the BC_a-interval in exponential families and call it ABC-internal, approximative bootstrap confidence interval.

3.4 Bootstrap Hypothesis Tests

In Example 3.2, a bootstrap confidence interval was used for testing H_0: $\theta = 1$, 1 lies in the interval, and therefore, we concluded that there is no evidence against H_0. In this section, we present various bootstrap approaches for approximating a p-value. One approach is to approximate the p-value by using the Monte Carlo method. The problem is that we need to sample from the null distribution, but we do not know where the samples are coming from. They can be from H_0 or H_1.

3.4 Bootstrap Hypothesis Tests

The testing setup is given as follows. We suppose $\mathbf{X} = (X_1, \ldots, X_n)$ is an i.i.d. sample from $X \sim P_\vartheta$ and we are interested in testing $\theta = g(\vartheta)$

$$H_0 : \theta(F) = \theta_0.$$

A test statistic $T = T(X_1, \ldots, X_n)$ with distribution P_ϑ^T is applied and the value is $t_{obs} = T(x_1, \ldots, x_n)$ is observed. The p-value is given by

$$p = \int I_{A_{obs}}(t)\, dP_0^T,$$

where $A_{obs} = A(t_{obs})$ describes the region in the direction of the extreme values, $I_A(t)$ is the indicator function, and P_0^T is the distribution of the test statistic T under H_0. The distribution P_0^T is called a null distribution. For small p-values we conclude that we have strong evidence against H_0, alternatively, we reject H_0 for p-values smaller than a given α.

If the null distribution P_0^T is known, the p-value is an integral, which then can be approximated by Monte Carlo methods. These tests are called Monte Carlo tests.

Bootstrap ideas are applied in cases where P_0^T is not completely given. We will study the behavior of T under H_0 and evaluate the case where the data are taken under the alternative. Thus, the main problem is:

> The bootstrap sampling must be performed under the null hypothesis.

In the subsequent section, we will present approaches for solving this problem.

3.4.1 Parametric Bootstrap Hypothesis Test

We assume that the distribution P_ϑ depends on the parameter of interest θ and a nuisance parameter λ, that is $\vartheta = (\theta, \lambda)$. The distribution of the test statistic $T = T(X_1, \ldots, X_n)$ depends on λ and θ. So, the null distribution $P_{(\theta_0, \lambda)}^T$ is not completely known. The p-value, however, can be determined by using a parametric bootstrap procedure.

Algorithm 3.3 Parametric Bootstrap Hypothesis Test

1. Estimate $\widehat{\lambda} = \lambda(x_1, \ldots, x_n)$.

2. Draw B random samples of size n from $P_{(\theta_0, \widehat{\lambda})}$:

$$X_{i,j}^* = x_{i,j}^*,\ X_{i,j}^* \sim P_{(\theta_0, \widehat{\lambda})},\ i = 1, \ldots, n,\ j = 1, \ldots, B.$$

Note that the samples are drawn under the hypothesis H_0.

3. Calculate the test statistics for each bootstrap sample

$$T_{(j)}^* = T\left(x_{1,j}^*, \ldots, x_{n,j}^*\right),\ j = 1, \ldots, B.$$

4. Approximate the p-value by

$$p_{boot} = \frac{\#(j : T^*_{(j)} \in A_{obs})}{B}.$$

Example 3.12 (Parametric bootstrap hypothesis test) We consider a test problem with two independent samples. The first sample consists of i.i.d. random variables $X_1, \ldots, X_n$ from a mixture normal distribution:

$$X_i = U_{1i} Z_{1i} + (1 - U_{1i}) Z_{2i},$$

where $U_{1i} \sim Ber(\frac{1}{2})$, $Z_{1i} \sim N(0,1)$, $Z_{2i} \sim N(\mu,1)$ is mutually independent and i.i.d.. The second sample $(Y_1, \ldots, Y_n)$ comes from a different mixture normal distribution

$$Y_i = U_{2i} V_{1i} + (1 - U_{2i}) V_{2i}$$

where $U_{2i} \sim Ber(\frac{3}{4})$, $V_{2i} \sim N(0,1)$, $V_{1i} \sim N(\mu + \Delta, 1)$ mutually independent and i.i.d.. The unknown parameter $\vartheta = (\Delta, \mu)$ consists of the shift parameter of interest Δ and of the nuisance parameter μ. The test problem is

$$H_0 : \Delta = 0 \text{ versus } H_1 : \Delta > 0.$$

We suppose as an appropriate test statistic $T = 4\overline{Y} - 2\overline{X}$, because

$$\begin{aligned} ET &= 4 E U_2 E V_1 + 4(1 - E U_2) E V_2 - 2 E U_1 E Z_1 - 2(1 - E U_1) E Z_2 \\ &= 4(1 - \frac{3}{4})(\mu + \Delta) - 2\frac{1}{2}\mu = \Delta. \end{aligned}$$

The nuisance parameter μ can be estimated by $\widehat{\mu} = 2\overline{X}$. The p-value is defined as

$$p = P_0 (T > T_{obs}).$$

Further, we have from the data: $n = 20, \overline{Y} = 2.211, \overline{X} = 2.57$, thus $T_{obs} = 3.685$ and $\widehat{\mu} = 5.152$. We can generate 1000 samples under $P_{(0,\widehat{\mu})}$. 58 samples of 1000 have a value higher than the observed T_{obs}. The p-value is then 0.058. The test result is that there is weak evidence against H_0, see Figure 3.9 and Figure 3.10. □

3.4.2 Nonparametric Bootstrap Hypothesis Test

Suppose that the null distribution of the test statistic is "completely" unknown. In some cases, one can find a resampling procedure which incorporates the null hypotheses. For example, for a two-sample problem

$$(Y_1, \ldots, Y_n) \sim P^Y; \quad (X_1, \ldots, X_m) \sim P^X \qquad H_0 : P^Y = P^X.$$

3.4 Bootstrap Hypothesis Tests

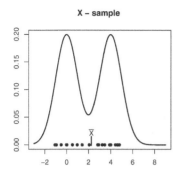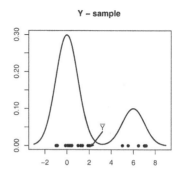

FIGURE 3.9: Simulated samples in Example 3.12.

Under H_0 the merged sample

$$\mathbf{Z} = (Z_1, \ldots, Z_{n+m}) = (Y_1, \ldots, Y_n, X_1, \ldots, X_m)$$

is an i.i.d. sample. We draw the bootstrap samples of size $n + m$ from $\widehat{F}_{n+m}$, the empirical distribution function of $\mathbf{Z}$.
Let us illustrate this approach with the help of the following example, see Figure 3.11.

Example 3.13 (Nonparametric bootstrap hypothesis test) The first sample consists of i.i.d. random variables from a mixture normal distribution and the second sample comes from the same mixture, but shifted by Δ. The null hypothesis is that both samples have the same distribution. We are interested in the two-sided test problem

$$H_0 : \Delta = 0 \text{ versus } H_1 : \Delta \neq 0.$$

We choose the test statistic

$$T(\mathbf{X}, \mathbf{X}) = |\bar{X} - \bar{Y}|.$$

From the data set, we have $T_{obs} = 2.34$, see Figure 3.11. The bootstrap samples are drawn by sampling with replacement from the merged sample. The number of bootstrap replications $T^*_{(j)}, j = 1, \ldots, 100$ which are less than T_{obs} is 20, compare the following R code. This has an approximated p-value of 0.02, thus we conclude that the distributions are significantly different. □

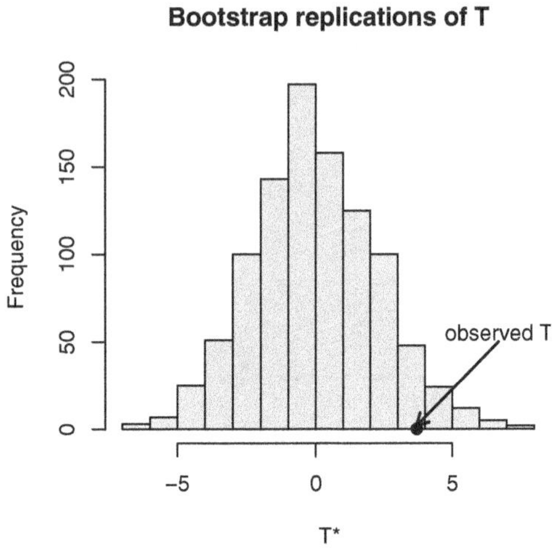

Bootstrap replications of T

FIGURE 3.10: Bootstrap p-value in Example 3.12.

R Code 3.4.17. Nonparametric bootstrap hypothesis test, Example 3.13.

```
T <- abs(mean(X)-mean(Y)) ### observed test statistic
Z <- c(X,Y)
TB <- rep(0,1000)
for (j in 1:1000){
    ZB <- sample(Z,replace=T);
    TB[j] <- abs(mean(ZB[1:10])-mean(ZB[11:20]))}
length(TB[T<TB])/1000 ### approximated p-value
```

3.4.3 Advanced Bootstrap Hypothesis Tests

There are more approaches to solve the problem of "sampling under H_0".

- A pivotal test statistics $T(\mathbf{X}, \theta)$ is used. The distribution of $T(\mathbf{X}, \theta)$ is indepedent of the parameter. We also recommend the prepivoting method of Beran (1988).

- Exploring a resampling strategy, which involves H_0 also in case when the data do not come from H_0. This can be done by using weighted bootstrap and a tilting method, see Davison and Hinkley (1997), p. 166.

3.5 Bootstrap in Regression

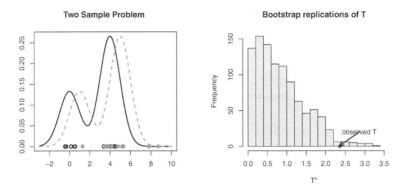

FIGURE 3.11: Simulated samples and bootstrapped test statistic in Example 3.13.

3.5 Bootstrap in Regression

In regression models, the data structure is more complex than in the i.i.d. case; therefore, a new approach to bootstrap sampling is needed. In this section, we present three different sampling strategies. For further reading, we recommend Davison and Hinkley (1997) and Freedman (1981).

3.5.1 Model-Based Bootstrap

Suppose a fixed design linear model:

$$Y = X\beta + \varepsilon; \quad E\varepsilon = 0; \quad \varepsilon = (\varepsilon_1, \ldots, \varepsilon_n)^T \quad \text{i.i.d. from } P,$$

where Y is the n-dimensional column vector of observations $y_1 \ldots, y_n$ and X is the $n \times p$ dimensional matrix of fixed design points, β is p dimensional column vector of the unknown parameter of interest. We always assume that the design matrix X has full rank. The observations $y_1, \ldots, y_n$ are independent but not identically distributed. The errors $\varepsilon_1, \ldots, \varepsilon_n$ are i.i.d., but they are not observed. The trick of bootstrap sampling is to estimate errors by the residuals, and then apply basic bootstrap to the residuals. Let us explain it in detail. The residuals are

$$\widehat{\varepsilon} = Y - X\widehat{\beta} = (\widehat{\varepsilon}_1, \ldots, \widehat{\varepsilon}_n)^T.$$

Using the least squares estimator $\widehat{\beta}$ we get

$$\widehat{\beta} = \left(X^T X\right)^{-1} X^T Y; \quad \widehat{\varepsilon} = Y - X\widehat{\beta} = (I - P)Y, \quad P = X(X^T X)^{-1} X^T.$$

Before sampling from the residuals we have to make sure that the empirical distribution function over the residuals has the expectation zero. Thus, a normalizing step is needed. We resample from

$$\widetilde{\varepsilon} = (\widehat{\varepsilon}_1 - \widehat{\mu}_n, \ldots, \widehat{\varepsilon}_n - \widehat{\mu}_n)^T \text{ with } \widehat{\mu}_n = \frac{1}{n}\sum_{i=1}^n \widehat{\varepsilon}_i.$$

Algorithm 3.4 Model-based Bootstrapping
Suppose data (X, Y).
Carry out a linear regression.

1. Normalize the residuals.

2. Draw the bootstrap sample ε^* from the empirical distribution function on $\widetilde{\varepsilon}$.

3. Calculate: $Y^* = X\widehat{\beta} + \varepsilon^*$.

4. Bootstrapped least squares estimator:
$\widehat{\beta}^* = (X^T X)^{-1} X^T Y^*$.

Note that the design matrix X remains unchanged. It is also possible to use other estimates for β than the least squares one.

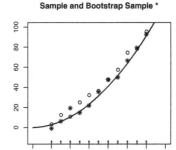
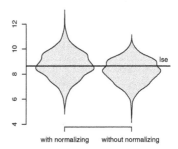

FIGURE 3.12: The effect of normalizing the residuals in Example 3.14.

For illustration, see the example below which shows the necessities of normalizing the residual.

Example 3.14 (Model-based bootstrap) Suppose the following special setting
$$y_i = \beta_1 x_i + \beta_2 x_i^2 + \varepsilon_i, \quad i = 1, \ldots, n.$$
The $x_1, \ldots, x_n$ are the equidistant design points. The design matrix X consists of one column with x_i as entries, and one column with x_i^2 as entries. The

3.5 Bootstrap in Regression

residuals are $\widehat{\varepsilon}_i = y_i - \widehat{\beta}_1 x_i + \widehat{\beta}_2 x_i^2$. It holds

$$\frac{1}{n}\sum_{i=1}^n \widehat{\varepsilon}_i = \mathbf{1}_n(I-P)\varepsilon,$$

where $\mathbf{1}_n = (1,\ldots,1)^T$ and $\varepsilon = (\varepsilon_1,\ldots \varepsilon_n)^T$. Under the design above $\mathbf{1}_n(I-P) \neq 0$. In Figure 3.12, on the left the bootstrap sample is plotted, which is

$$y_i^* = \widehat{\beta}_1 x_i + \widehat{\beta}_2 x_i^2 + \widetilde{\varepsilon}_i^*.$$

The $\widetilde{\varepsilon}_i^*, i = 1,\ldots,n$ are sampled with replacement from

$$\widetilde{\varepsilon}_i = \widehat{\varepsilon}_i - \frac{1}{n}\sum_{i=1}^n \widehat{\varepsilon}_i, i = 1,\ldots,n.$$

The violin plot in Figure 3.12 demonstrates the difference when the normalizing step is omitted. □

R Code 3.5.18. Model-based bootstrap, Example 3.14.

```
x2 <- x*x
M <- lm(y~x+x2-1) ### regression without intercept
R <- M$resid-mean(M$resid) ### normalizing residuals
B <- 1000
B2 <- rep(0,B)
for (j in 1:B){
    yb <- M$fit+sample(R,replace=T)
    Mb <- lm(yb~x+x2-1)
    B2[j] <- Mb$coef[2]}
```

3.5.2 Parametric Bootstrap Regression

This method is similar to the model-based bootstrap; the difference is that the residuals are generated by a parametric distribution. For this method, it is not necessary to have the true class of error distribution; however, it is important that the expectation is zero.

Algorithm 3.5 Parametric Bootstrap Regression
Suppose data (X, Y).
Carry out parametric linear regression.

1. Draw the bootstrap sample ε^* from some distribution P, with expectation zero.
2. Calculate: $Y^* = X\widehat{\beta} + \varepsilon^*$.

---

3. Bootstrapped least squares estimator:
$$\widehat{\beta}^* = \left(X^T X\right)^{-1} X^T Y^*.$$

We continue with Example 3.14.

Example 3.15 (Parametric bootstrap regression) The bootstrap sample is now
$$y_i^* = \widehat{\beta}_1 x_i + \widehat{\beta}_2 x_i^2 + \varepsilon_i^*.$$
The $\varepsilon_i^*, i = 1, \ldots, n$ are sampled from $N(0, 1)$. The violin plot in Figure 3.13 demonstrates the difference when the error distribution is changed. In the example here, the generating error distributions are chosen such that their variances are equal. □

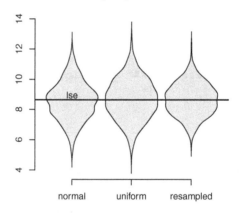

FIGURE 3.13: Example 3.15.

3.5.3 Casewise Bootstrap in Correlation Model

The model here is different from parametric bootstrap regression. The independent X-variables are random. We are back to the situation of an i.i.d. sample. We assume $(y_i, x_i), i = 1, \ldots, n$, are i.i.d. with
$$y_i = x_i^T \beta + \varepsilon_i \ ; \ E\varepsilon = 0; \ \varepsilon = (\varepsilon_1, \ldots, \varepsilon_n)^T \text{ i.i.d. from } P,$$

3.5 Bootstrap in Regression

where $x_i = (x_{i1}, \ldots, x_{ip})^T$ and β is p dimensional. The $n \times p$ X matrix consists of the x_{ij}. We resample the vectors (y_i^*, x_i^*) from (y_i, x_i) $i = 1, \ldots, n$ and calculate the bootstrap replications of

$$\widehat{\beta} = \left(X^T X\right)^{-1} X^T Y \text{ by } \widehat{\beta}^* = \left(X^{*T} X^*\right)^{-1} X^{*T} Y^*.$$

This method is also called casewise bootstrapping, see R Code 3.5.19. It is different from parametric boostrapping in that the X-variables are samples. Note that it is absolutely essential that the pairs (y_i, x_i) $i = 1, \ldots, n$ remain connected.

Remark: (Bootstrap hypothesis test) Suppose we want to test $H_0 : \beta = 0$. Under H_0, X and Y are independent. In that case, to sample under the null hypothesis means to sample X and Y independently.

R Code 3.5.19. Casewise resampling

```
function(x,y){
    n <- length(x)
    data <- data.frame(x,y)
    for(i in 1:n){
        u <- ceiling(runif(1,0,n))
        data[i,1] <- x[u]
        data[i,2] <- y[u]}
    return(data)}
```

In the following example, we demonstrate pairwise sampling for the historical data set of Galton.

Example 3.16 (Galton's data) Francis Galton (1822-1911) was interested in the estimate of heredity. His famous data set from 1885 is a table of the height y_i of the child i in inches, and the height of its parent x_i (the average of the father's height and 1.08 times of the mother's height). All heights of the daughters are multiplied by 1.08. The data set contains 205 different parents and 928 children. This data set is also known as one of the first examples of a regression analysis. We have

$$y_i = \alpha + \beta x_i + \varepsilon_i, \ i = 1, \ldots, n = 928,$$

$$y_i \in \{61.7, 62.2, 62.7, \ldots, 73.2, 73.7\} \, , \ x_i \in \{64, 64.5, 65, \ldots, 72.5, 73\}$$

Because of the tabulation, the heights are discretized and appear more than once. Therefore, we resample from a discrete distribution, see the R Code 3.5.20 and Figure 3.14. Let us study the hypotheses that the height of the child is positively correlated with the height of the parent. This means that we are interested in testing

$$H_0 : \beta = 0 \text{ versus } H_1 : \beta > 0.$$

In contrast to Galton, we apply the least squares estimator

$$\widehat{\beta} = \frac{\sum_{i=1}^{n}(x_i - \overline{x})(y_i - \overline{y})}{\sum_{i=1}^{n}(x_i - \overline{x})^2},$$

and then apply bootstrap. Figure 3.15 contains the histogram of the bootstrap replications $\widehat{\beta}^*_{(j)}, j = 1, \ldots, 1000$. The lower confidence bound is 0.576. We conclude that the heights are positively correlated. □

R Code 3.5.20. Casewise bootstrap for Galton's table, Example 3.16.

```
### 2D histogram, M matrix with the table values
library(lattice)
library(latticeExtra)
cloud(M,xlab="parent",ylab="child",zlab="Freq",zoom=1.2,
par.settings=list(box.3d=list(col="transparent"))
,col.facet=grey(0.9),screen=list(z=40,x=-25),
panel.3d.cloud=panel.3dbars,panel.aspect=1.3,
aspect=c(1,1),xbase=0.6,ybase=0.6)
### bootstrap
library(UsingR)
data(galton)
attach(galton)
M <- lm(child~parent)
M$coeff[2]
```

3.6 Bootstrap for Time Series

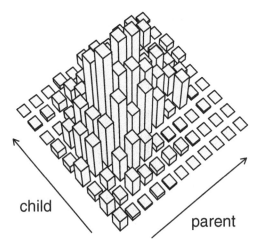

FIGURE 3.14: Galton's table in Example 3.16.

```
boot <- function(x,y,B){
    BB <- rep(0,B)
    n <- length(y)
    for( j in 1:B){
        u <- sample(1:n,replace=T)
        M <- lm(y[u]~x[u])
        BB[j] <- M$coeff[2]}
    return(BB)}
Bound <- sort(boot(parent,child,1000))[50]
```

3.6 Bootstrap for Time Series

In Example 3.3, we already presented a model-based bootstrap sample for an AR(1) model. Here, we explain the most common procedures for bootstrapping time series: blockwise resampling by Künsch (1989) and the stationary bootstrap by Politis and Romano (1994).

The bootstrap sample should have the same properties as the observed data. In a case where the data are i.i.d. samples, we intend to generate a conditional i.i.d. bootstrap sample. In this section, we consider time series and we want that the bootstrapped time series should keep their dependence structure as much as possible. Künsch (1989) proposed a blockwise resampling method. The idea behind this is that the blocks behave like an i.i.d. sample.

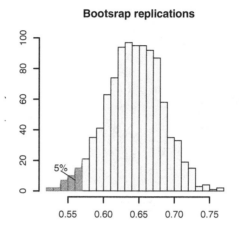

Bootsrap replications

FIGURE 3.15: Casewise bootstrap replications in Example 3.16.

Algorithm 3.6 Blockwise Bootstrapping

1. Propose a fixed block length l. Suppose $n = Kl$.
2. Divide the time series into K blocks of length l. Count the blocks with index $k = 1, \ldots, K$. Then
$$X = (X_1, \ldots, X_n) = (B_1, \ldots, B_K)$$
with
$$B_k = (X_{(k-1)l+1}, \ldots, X_{kl}).$$
3. Sample with drawback $k_1^*, \ldots, k_K^*$ from $1, \ldots, K$.
4. The bootstrapped time series is then
$$X^* = (B_{k_1^*}, \ldots, B_{k_K^*}).$$

This procedure is implemented in R in the package `boot`.

R Code 3.6.21. Blockwise bootstrap, Example 3.3.

```
### simulate an AR(1) time series
T <- arima.sim(n = 1000, list(ar = c(0.8)), sd = sqrt(0.1796))
acf(TS) ### autocorrelation  for the choice of the block length

library(boot)
stat <- function(TS){TS}
```

3.6 Bootstrap for Time Series

```
### one bootstrapped time series, block length=5
TB <- tsboot(TS,stat,sim="fixed",R=1,l=5,n.sim=100,orig.t=FALSE)
### estimator in AR(1)
stat2 <- function(T){arima(T,c(1,0,0),include.mean=F)$coef[1]}
TB2 <- tsboot(TS,stat2,sim="fixed",R=1000,l=5,
n.sim=100,orig.t=TRUE)
TB2$t ### 1000 bootstrap replications of the estimator
```

Blockwise resampling does not generate a stationary time series. Politis and Romano (1994) introduced a blockwise resampling where the block length is geometrically distributed. They call their procedure *the stationary bootstrap* and state that the generated bootstrap time series is stationary, conditional on the original time series. Let $B_{i,b}$ be the block starting with X_i and of length b:

$$B_{i,b} = (X_i, \ldots, X_{i+b-1}).$$

Algorithm 3.7 The Stationary Bootstrap

1. Choose a tuning parameter $p, 0 < p < 1$.

2. Generate independently $L_1, L_2 \ldots$ from a geometric distribution with parameter p.

3. Draw independently $I_1, I_2, \ldots$ uniformly distributed on $1, \ldots, n$.

4. The bootstrapped time series of length N is then

$$X^* = (B_{I_1, L_1}, B_{I_2, L_2}, \ldots).$$

R Code 3.6.22. Stationary bootstrap, Example 3.3.

```
### simulate an AR(1) time series
T <- arima.sim(n = 1000, list(ar = c(0.8)), sd = sqrt(0.1796))
acf(TS) ### autocorrelation for the choice of the block length

library(boot)
stat <- function(TS){TS}
### one bootstrapped time series block length geometrical
### distributed with EX=1
TB <- tsboot(TS,stat,sim="geom",R=1,l=5,n.sim=100,orig.t=FALSE)
### estimator in AR(1)
stat2 <- function(T){arima(T,c(1,0,0),include.mean=F)$coef[1]}
TB2 <- tsboot(TS,stat2,sim="geom",R=1000,l=5
,n.sim=100,orig.t=TRUE)
TB2$t ### 1000 bootstrap replications of the estimator
```

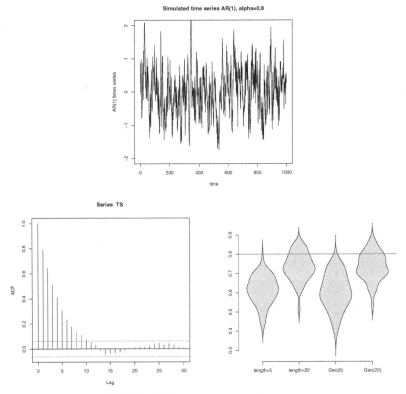

FIGURE 3.16: Bootstrap for time series.

See Figure 3.16 for comparison of the different methods.

3.7 Problems

Spatial data set: Data is quoted from Efron and Tibshirani (1993). Twenty-six neurologically impaired children have taken two sets of spatial perception, called "A" and "B".

i	1	2	3	4	5	6	7	8	9	10	11	12	13
A	48	36	20	29	42	42	20	42	22	41	45	14	6
B	42	33	16	39	38	36	15	33	20	43	34	22	7

i	14	15	16	17	18	19	20	21	22	23	24	25	26
A	0	33	28	34	4	32	24	47	41	24	26	30	41
B	15	34	29	41	13	38	25	27	41	28	14	28	40

3.7 Problems

1. Use the spatial data set. We are interested in the variance of the first test A: $\theta = Var(A)$.

   (a) Calculate the plug-in-estimator $\hat{\theta}$ for θ.
   (b) Generate nonparametric bootstrap replications of $\hat{\theta}$. Plot the histogram. Calculate a 95%–nonparametric basic bootstrap interval for θ.
   (c) Generate parametric bootstrap replications of $\hat{\theta}$. Plot the histogram. Calculate a 95%–parametric basic bootstrap interval for θ.
   (d) Derive a 95%–studentized bootstrap interval for θ.
   (e) Compare the results.

2. Use the spatial data set. Set: $A_i = a + bB_i + \varepsilon_i$

   (a) Generate bootstrap replications for the least squares estimator of b; by casewise resampling. Plot the histogram.
   (b) Derive a percentile bootstrap confidence interval for b.
   (c) Derive a bootstrap hypothesis test for $H: b = 1$.

3. Consider an i.i.d. sample $(x_1, ..., x_n)$ from $N(0, \theta)$ and $\tilde{\theta} = \frac{1}{n}\sum_{i=1}^n x_i^2$.

   (a) Describe a parametric bootstrap sampling strategy and a nonparametric bootstrap sampling strategy.
   (b) Which conditional distribution P^* has $\tilde{\theta}^*$ calculated from one parametric bootstrap sample $(x_1^*, .., x_n^*)$?
   (c) Are the parametric bootstrap replications $\tilde{\theta}_{(j)}^*$, $j = 1, ..., B$ independent (under the unconditional distribution!) distributed? (Hint: $Cov(y, y) = E_z(Cov(x, y)/z) + (Cov(Ex/z, Ey/z))$).

4. Consider an i.i.d. sample $X_1, \ldots, X_n$ from a *Pareto* distribution

   $$p(x) = \frac{\alpha k^\alpha}{x^{\alpha+1}}, \quad x > k > 0, \ \alpha > 2.$$

   Note $EX = \frac{\alpha k}{\alpha-1}, VarX = \frac{\alpha k^2}{(\alpha-2)(\alpha-1)^2}$. Set $k = 2$.

   (a) Suppose a plug-in estimator for $\theta = \frac{\alpha}{\alpha-1}$.
   (b) Construct a nonparametric basic bootstrap confidence interval for θ.
   (c) Construct a nonparametric bootstrap confidence interval for α using the nonparametric basic bootstrap interval for θ.

5. Consider the Stroke example, Example 3.2. We are interested in $\theta = \frac{p_1}{p_2}$ where p_1 is the success probability of the treatment group and p_2 is the success probability of the control group.

(a) Describe a parametric bootstrap sampling strategy and a non-parametric bootstrap sampling strategy. Compare both strategies.

(b) Perform a bootstrap hypothesis test for $H_0: \theta = 1$ versus $H_1: \theta \neq 1$.

6. Consider the following regression model

$$y_i = \beta x_i + \varepsilon_i; \; i = 1, \ldots, n, \; x_i = \frac{i}{n}, \; \varepsilon_i \text{ i.i.d. } N(0,1).$$

(a) Calculate the least squares estimator $\widehat{\beta}$ for β. Which distribution has $\widehat{\beta}$?

(b) Consider the residuals

$$e_i = y_i - \widehat{y}_i, \; \widehat{y}_i = x_i \widehat{\beta}.$$

Resample conditional independently $(e_1^*, \ldots, e_n^*)$ from $\{e_1, \ldots, e_n\}$ such that

$$P(e_l^* = e_j \mid y_1, \ldots, y_n) = \frac{1}{n}, \; i = 1, \ldots, n, \; j = 1, \ldots, n.$$

Set

$$y_i^* = \widehat{\beta} x_i + e_i^*, \; i = 1, \ldots, n.$$

Which conditional distribution has a bootstrap replication $\widehat{\beta}^*$ of $\widehat{\beta}$? Calculate the expectation and variance of this conditional distribution.

(c) Consider a sample of bootstrap replications $\widehat{\beta}^*_{(1)}, \ldots, \widehat{\beta}^*_{(B)}$. Is it possible to approximate the distribution of $\widehat{\beta}$ by $\widehat{\beta}^*_{(1)}, \ldots, \widehat{\beta}^*_{(B)}$? Why or why not?

7. Consider the following regression model

$$y_i = \beta x_i + \varepsilon_i; \; i = 1, 2, 3,$$

where x_i are i.i.d. with $P(x_i = k) = \frac{1}{4}, k \in \{-2, -1, 1, 2\}$ and ε_i are i.i.d. with $P(\varepsilon_i = k) = \frac{1}{2}, k \in \{-1, 1\}$; ε_i and x_i are mutually independent.

(a) Calculate the least squares estimator $\widehat{\beta}$ for β. Which distribution has $\widehat{\beta}$?

(b) Resample casewise from $\{(x_1, y_1), (x_2, y_2), (x_3, y_3)\}$. Calculate the distribution (unconditional) of a bootstrap replication $\widehat{\beta}^*$ of $\widehat{\beta}$.
Calculate the expectation and variance of this distribution.

3.7 Problems

(c) Consider a sample of bootstrap replications $\widehat{\beta}^*_{(1)}, \ldots, \widehat{\beta}^*_{(B)}$. Is it possible to approximate the distribution of $\widehat{\beta}$ by $\widehat{\beta}^*_{(1)}, \ldots, \widehat{\beta}^*_{(B)}$?

8. Compare the examination results of students in autumn 2012 and autumn 2013. The table below shows the number of students that reached the respective marks.

	A	B	C	sample size
autumn 12	x_1	x_2	x_3	n_1
autumn 13	y_1	y_2	y_3	n_2

Let p_1 be the probability to get a mark A or B in autumn 12, and let p_2 be the probability to get a mark A or B in autumn 13. We are interested in $\theta = \frac{p_1}{p_2}$.

(a) Propose an estimator for the probabilities p_1, p_2, and θ.

(b) Write the main steps of a bootstrap method for estimating the distribution of $\theta = \frac{p_1}{p_2}$.

(c) Write the main steps for performing a bootstrap test for $H_0 : \theta = 1$ versus $H_1 : \theta < 1$.

9. The following case study is quoted from the textbook by Weiss (2015): "The U.S. National Center for Health Statistics compiles data of stay by patients in short-term hospitals and publishes its findings in *Vital and Health Statistics*. Independent samples of 39 male patients and 35 female patients gave the following data on length of stay in days."

```
Male:    4    4   12   18    9    6   12   10    3
         6   15    7    3   13    1    2   10   13    5
         7    1   23    9    2    1   17    2   24   11
        14    6    2    1    8    1    3   19    3    1

Female: 14    7   15    1   12    1    3    7   21
         4    1    5    4    4    3    5   18   12    5
         1    7    7    2   15    4    9   10    7    3
         6    5    9    6    2   14
```

(a) Perform a Shapiro test for both samples.

(b) Interpret the p-values of the Shapiro tests.

(c) 5% significance level test is required for comparing the length of stay. Which test statistics do you propose?

(d) Write the main steps for performing a bootstrap test.

(e) Write a short R code for the procedure in (d).

4
Simulation-Based Methods

In this chapter, we introduce methods that are based on the embedding principle. Assume we have a data set that belongs to a complicated model. If we would had more observations, unconstrained variables without missing values, and no hidden variables, we could find a big and "easier" model, and apply "easier" methods.

The trick of embedding-based methods is to find a big model and to simulate additional observations. Then we combine the original data set with the simulated observations and apply methods of the big model. It is just a change of the perspective, as in Figures 4.1 and 4.2.

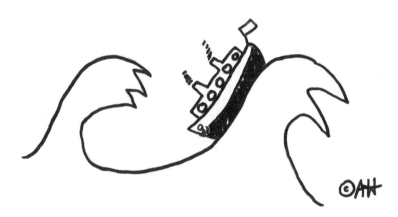

FIGURE 4.1: The observed environment.

FIGURE 4.2: Change the perspective!

4.1 EM Algorithm

The EM (expectation and maximization) algorithm is the oldest and maybe one of the most common methods based on the embedding principle. It was introduced in 1977 by Dempster et al. (1977). For more details regarding this algorithm, see Chapters 10 and 12 of Lange (1999).

In the original formulation of the EM algorithm, there is no simulation step; instead, a projection method is proposed. The setup is formulated as follows. Assume a data set includes the following observations:

$$\mathbf{Y} = (Y_1, \ldots, Y_m)$$

from the **observed model** with density $g(\mathbf{y} \mid \theta)$. We wanted to calculate the maximum likelihood estimator of θ. The observations are interpreted as a transformation

$$\mathbf{Y} = T(\mathbf{X}), \tag{4.1}$$

where

$$\mathbf{X} = (X_1, \ldots, X_n)$$

belongs to a **complete model** with density $f(\mathbf{x} \mid \theta)$. The transformation in (4.1) means that the observed data $\mathbf{y}$ are augmented to the data set $\mathbf{x}$. The trick is to use the projection (conditional expectation) of the log-likelihood function $\ln f(\mathbf{x} \mid \theta)$ as an estimation criterion. The conditional mean is a projection of the likelihood function from the complete model on to the observed

4.1 EM Algorithm

model. In the original article, it was called the surrogate function and is defined by

$$Q(\theta \mid \theta') = E(\ln f(\mathbf{X} \mid \theta) \mid T(\mathbf{X}), \theta').$$

We have two parameters: θ is a free parameter in the likelihood function, and θ' is the parameter of the underlying distribution. Here, it is assumed that the transformation is insufficient. The case of a sufficient statistic T is uninteresting because, in that case, minimizing of the surrogate function is the same as minimizing the likelihood function $l(\theta) = \ln g(\mathbf{y} \mid \theta)$. For insufficient transformations, the conditional expectation of $\mathbf{X}$ given $T(\mathbf{X})$ depends on the true underlying parameter θ_{true}. The EM algorithm is an iterative procedure for solving

$$\arg\max_\theta Q(\theta \mid \theta_{true}).$$

Algorithm 4.1 EM Algorithm

Given the current state $\theta^{(k)}$.

1. E-step (Expectation): Calculate $Q(\theta \mid \theta^{(k)})$.

2. M-step (Maximization):

$$\theta^{(k+1)} = \arg\max_\theta Q(\theta \mid \theta^{(k)}).$$

Why does EM work?

First, we prove the following theorem, the information inequality. This inequality is the main argument for the maximum likelihood principle.

Theorem 4.1 *Let $f > 0$ a.s, and $g > 0$ a.s. be two densities, then*

$$E_f \ln f \geq E_f \ln g.$$

Proof:
According to the Jensen inequality, let W be a random variable and let $h(w)$ be a convex function, then

$$Eh(W) \geq h(E(W))$$

provided both expectations exist. Note that the function $-\ln(w)$ is strictly convex. Thus

$$E_f \ln f - E_f \ln g = -E_f \ln\left(\frac{g}{f}\right) \geq -\ln E_f\left(\frac{g}{f}\right).$$

Furthermore,
$$E_f\left(\frac{g}{f}\right) = \int g\,dx = 1.$$

We get
$$E_f \ln f - E_f \ln g \geq -\ln(1) = 0.$$

□

Now, we compare the surrogate function with the log-likelihood of the observed model
$$l(\theta) = \ln g(\mathbf{y} \mid \theta).$$

Theorem 4.2 *It holds*
$$Q(\theta' \mid \theta') - l(\theta') \geq Q(\theta \mid \theta') - l(\theta). \tag{4.2}$$

Proof: We have
$$\begin{aligned} Q(\theta \mid \theta') - l(\theta) &= E \ln(f(\mathbf{X} \mid \theta) \mid \mathbf{Y} = \mathbf{y}, \theta') - \ln g(\mathbf{y} \mid \theta) \\ &= E \ln(\frac{f(\mathbf{X} \mid \theta)}{g(\mathbf{y} \mid \theta)} \mid \mathbf{Y} = \mathbf{y}, \theta'). \end{aligned}$$

Note
$$\frac{f(\mathbf{X} \mid \theta)}{g(\mathbf{y} \mid \theta)}$$
is the density of the conditional distribution of $\mathbf{X} \mid (\mathbf{Y} = \mathbf{y}, \theta)$. We apply the information inequality and obtain
$$Q(\theta \mid \theta') - l(\theta) \leq E \ln(\frac{f(\mathbf{X} \mid \theta')}{g(\mathbf{y} \mid \theta')} \mid \mathbf{Y} = \mathbf{y}, \theta') = Q(\theta' \mid \theta') - l(\theta'). \tag{4.3}$$

□

The inequality (4.2) says that a maximizing of the surrogate function implies an increase of the likelihood function $l(\theta)$ because for
$$\theta^{(k+1)} = \arg\max_\theta Q\left(\theta \mid \theta^{(k)}\right)$$
it holds
$$\begin{aligned} l(\theta^{(k+1)}) &\geq Q\left(\theta^{(k+1)} \mid \theta^{(k)}\right) + l(\theta^{(k)}) - Q\left(\theta^{(k)} \mid \theta^{(k)}\right) \\ &\geq l(\theta^{(k)}). \end{aligned}$$

In the paper from Dempster et al. (1977), the following example was discussed.

4.1 EM Algorithm

Example 4.1 (EM algorithm: genetic linkage model) Given data

$$\mathbf{y} = (y_1, y_2, y_3, y_4) = (125, 18, 20, 34)$$

from a four-category multinomial distribution

$$\mathbf{Y} \sim multinomial(n, p_1, p_2, p_3, p_4),$$

with

p_1	p_2	p_3	p_4
$\frac{1}{2} + \frac{1}{4}p$	$\frac{1}{4}(1-p)$	$\frac{1}{4}(1-p)$	$\frac{1}{4}p$

We represent the data $\mathbf{y}$ as incomplete data from a five-category multinomial population:

$$y_1 = x_1 + x_2, \ y_2 = x_3, \ y_3 = x_4, \ y_4 = x_5, \quad (4.4)$$

with

$$\mathbf{X} \sim multinomial(n, p_1, p_2, p_3, p_4, p_5), \quad (4.5)$$

where

p_1	p_2	p_3	p_4	p_5
$\frac{1}{2}$	$\frac{1}{4}p$	$\frac{1}{4}(1-p)$	$\frac{1}{4}(1-p)$	$\frac{1}{4}p$

(4.6)

The conditional distribution of $(X_1, X_2) \mid Y_1 = y_1$ is

$$(X_1, X_2) \mid Y_1 = y_1 \sim multinomial(y_1, \frac{p_1}{p_1 + p_2}, \frac{p_2}{p_1 + p_2}).$$

Especially

$$X_1 \mid Y_1 = y_1 \sim Bin(y_1, \frac{p_1}{p_1 + p_2}), \quad X_2 \mid Y_1 = y_1 \sim Bin(y_1, \frac{p_2}{p_1 + p_2}).$$

We are interested in a maximum likelihood estimation of p. Then, under this setup, we can write the following EM algorithm.

EM Algorithm: Genetic Linkage Example

Given the current state $p^{(k)}$:

1. (E) Estimate the "missing" observation x_2 by the conditional expectations.

$$x_2^{(k)} = E(X_2 \mid Y_1 = y_1, p^{(k)}) = y_1 \frac{\frac{1}{4}p^{(k)}}{\frac{1}{2} + \frac{1}{4}p^{(k)}}.$$

2. (M) Calculate the maximum likelihood estimation $p^{(k+1)}$ in model (4.5) by:

$$p^{(k+1)} = \frac{x_2^{(k)} + y_4}{x_2^{(k)} + y_2 + y_3 + y_4}.$$

Let us compare the algorithm above with the general EM algorithm. First, we consider the more general setup of a multinomial model with additional parameterized cell probabilities

$$\mathbf{X} \sim multinomial(n, p_1(\theta), \ldots, p_K(\theta)).$$

In this model, the log-likelihood function up to an additive constant is

$$\sum_{j=1}^{K} x_j \ln(p_j(\theta)) = \sum_{j=1}^{K-1} x_j \ln(p_j(\theta)) + (n - \sum_{j=1}^{K-1} x_j) \ln(1 - \sum_{j=1}^{K-1} p_j(\theta))$$

and the surrogate function up to an additive constant is

$$\begin{aligned} Q\left(\theta \mid \theta^{(k)}\right) &= E\left(\sum_{j=1}^{K} x_j \ln(p_j(\theta)) \mid \mathbf{Y} = \mathbf{y}, \theta^{(k)}\right) \\ &= \sum_{j=1}^{K} \ln(p_j(\theta)) E(x_j \mid \mathbf{Y} = \mathbf{y}, \theta^{(k)}), \end{aligned}$$

which is just the maximum likelihood function with the estimated data

$$\hat{x}_j^{(k)} = E(x_j \mid \mathbf{Y} = \mathbf{y}, \theta^{(k)}).$$

Consider (4.6). $l'(\theta) = 0$ respects to

$$x_2 \frac{1}{p} - x_3 \frac{1}{1-p} - x_4 \frac{1}{1-p} + x_5 \frac{1}{p} = 0.$$

Note that x_1 is not included, because the probability $p_1 = \frac{1}{2}$ does not depend on the parameter of interest p. We get

$$\widehat{p}_{ML} = \frac{x_2 + x_5}{x_2 + x_3 + x_4 + x_5}.$$

In the observed model, the likelihood equation is more complicated to solve. We have

$$y_1 \frac{1}{\frac{1}{2} + \frac{1}{4}p} - (y_2 + y_3) \frac{1}{1-p} + y_4 \frac{1}{p} = 0.$$

Thus, using the MLE in the complete model with estimated observations $\hat{x}_2$ give us an easier expression. □

4.1 EM Algorithm

In the case of multinomial distributions with incomplete data, the calculations in the E-step and the M-step are carried out explicitly. Another important example is mixture distributions, which belong to models with a latent unobserved variable. Also, in this case, some of the calculations in the E-step and the M-step can be done explicitly.

Example 4.2 (EM algorithm: mixture distribution)
Assume an i.i.d. sample $\mathbf{Y} = (Y_1, \ldots, Y_n)$ from $Y \sim F$, a mixture distribution with density $g(y \mid \theta)$, $\theta = (\pi, \vartheta_1, \vartheta_2)$

$$g(y \mid \theta) = (1 - \pi) f_{1,\vartheta_1}(y) + \pi f_{1,\vartheta_1}(y), \ \pi \in [0, 1]. \tag{4.7}$$

Mixture distributed random variables have a stochastic decomposition:

$$\begin{aligned} Y &= \Delta \, Z_1 + (1 - \Delta) \, Z_2, \\ Z_1 &\sim F_1 \text{ with } f_1 = f_{1,\vartheta_1}, \quad Z_2 \sim F_2, \text{ with } f_2 = f_{2,\vartheta_2}, \text{ independent}, \\ \Delta &\sim Ber(\pi) \text{ independent of } Z_1 \text{ and } Z_2. \end{aligned}$$

The Bernoulli distributed random variable $\Delta \sim Ber(\pi)$ is latent and unobservable. The complete model is the following. Let

$$\mathbf{X} = (X_1, \ldots, X_n)$$

be an i.i.d. sample with

$$X = (Y, \Delta) \ \Delta \sim Ber(\pi), \ X \mid \Delta = 0 \sim F_1, \ X \mid \Delta = 1 \sim F_2.$$

Then

$$f(x \mid \theta) = f(y \mid \Delta, \theta) \, f(\Delta \mid \theta) = (f_{1,\vartheta_1}(y)(1-\pi))^{(1-\Delta)} (f_{2,\vartheta_2}(y)\pi)^{\Delta}.$$

Thus, the log-likelihood of the complete model is

$$\ln f(\mathbf{x} \mid \theta) = \sum_{i=1}^{n} (1 - \Delta_i) \ln (f_{1,\vartheta_1}(y_i)(1-\pi)) + \Delta_i \ln (f_{2,\vartheta_2}(y_i)\pi).$$

The observed model is $T(\mathbf{X}) = \mathbf{Y}$ with density (4.7). The surrogate function is

$$Q\left(\theta \mid \theta^{(k)}\right) = \sum_{i=1}^{n} (1 - \gamma_{i,k}) \ln (f_{1,\vartheta_1}(y_i)(1-\pi)) + \gamma_{i,k} \ln (f_{2,\vartheta_2}(y_i)\pi),$$

with $\gamma_{i,k} = \gamma(y_i, \theta^{(k)})$, where

$$\begin{aligned}
\gamma(y,\theta) &= E(\Delta \mid T(X) = y, \theta) \\
&= P(\Delta = 1 \mid T(X) = y, \theta) \\
&= \frac{P(y \mid \Delta = 1, \theta) P(\Delta = 1)}{P(y \mid \theta)} \\
&= \frac{f_{2,\vartheta_2}(y)\pi}{f_{1,\vartheta_1}(y)(1-\pi) + f_{2,\vartheta_2}(y)\pi}.
\end{aligned}$$

The $\gamma_{i,k}$ are called responsibilities and are interpreted as kth iteration of the estimates of the latent variable Δ_i.
For the calculation of

$$\theta^{(k+1)} = \arg\max_\theta Q\left(\theta \mid \theta^{(k)}\right)$$

we solve the normal equations

$$\frac{d}{d\vartheta_1} Q\left(\theta \mid \theta^{(k)}\right) = \sum_{i=1}^n (1 - \gamma_{i,k}) \frac{d}{d\vartheta_1} \ln(f_{1,\vartheta_1}(y_i)) = 0$$

$$\frac{d}{d\vartheta_2} Q\left(\theta \mid \theta^{(k)}\right) = \sum_{i=1}^n \gamma_{i,k} \frac{d}{d\vartheta_2} \ln(f_{2,\vartheta_2}(y_i)) = 0$$

and

$$\frac{d}{d\pi} Q\left(\theta \mid \theta^{(k)}\right) = \sum_{i=1}^n (1 - \gamma_{i,k}) \frac{-1}{1-\pi} + \gamma_{i,k} \frac{1}{\pi} = 0.$$

The last equation gives $\pi^{(k+1)} = \frac{1}{n} \sum_{i=1}^n \gamma_{i,k}$.
We now summarize the calculations and formulate the algorithm.

Algorithm 4.2 EM Algorithm: Mixture Distributions

Given the current state $\theta^{(k)} = \left(\pi^{(k)}, \vartheta_1^{(k)}, \vartheta_2^{(k)}\right)$.

1. Calculate the responsibilities for $i = 1, \ldots, n$

$$\gamma_{i,k} = \frac{f_{2,\vartheta_2^{(k)}}(y_i)\pi^{(k)}}{f_{1,\vartheta_1^{(k)}}(y_i)(1-\pi^{(k)}) + f_{2,\vartheta_2^{(k)}}(y_i)\pi^{(k)}}.$$

2. Find ϑ_1 such that

$$\sum_{i=1}^n (1 - \gamma_{i,k}) \frac{d}{d\vartheta_1} \ln(f_{1,\vartheta_1}(y_i)) = 0,$$

4.1 EM Algorithm

> and find ϑ_2 such that
>
> $$\sum_{i=1}^{n} \gamma_{i,k} \frac{d}{d\vartheta_2} \ln\left(f_{2,\vartheta_2}(y_i)\right) = 0.$$
>
> 3. Update $\theta^{(k+1)} = \left(\pi^{(k+1)}, \vartheta_1^{(k+1)}, \vartheta_2^{(k+1)}\right)$ with
>
> $$\pi^{(k+1)} = \frac{1}{n} \sum_{i=1}^{n} \gamma_{i,k}, \quad \vartheta_1^{(k+1)} = \vartheta_1, \quad \vartheta_2^{(k+1)} = \vartheta_2.$$

Consider the example of the mixture of two normals, $F_1 = N(\mu_1, \sigma_1^2)$, $F_2 = N(\mu_2, \sigma_2^2)$, Figure 4.3. Then

$$\mu_1^{(k+1)} = \frac{\sum_{i=1}^{n}(1-\gamma_{i,k})y_i}{\sum_{i=1}^{n}(1-\gamma_{i,k})}, \quad \mu_2^{(k+1)} = \frac{\sum_{i=1}^{n}\gamma_{i,k}y_i}{\sum_{i=1}^{n}\gamma_{i,k}}, \tag{4.8}$$

and

$$\sigma_{1,k+1}^2 = \frac{\sum_{i=1}^{n}(1-\gamma_{i,k})(y_i-\mu_1^{(k+1)})^2}{\sum_{i=1}^{n}(1-\gamma_{i,k})}, \quad \sigma_{2,k+1}^2 = \frac{\sum_{i=1}^{n}\gamma_{i,k}(y_i-\mu_2^{(k+1)})^2}{\sum_{i=1}^{n}\gamma_{i,k}}. \tag{4.9}$$

□

R Code 4.1.23. EM Algorithm for normal mixture

In this example, for simplicity, we assume that π is known.

```
data <- Z ### data simulated from two normal mixtures
pi <- 0.3
r <- rep(NA,100) ### responsibilities
Mu1 <- rep(0,11)
### series of estimate of the first expectation
Mu2 <- rep(0,11)
### series of estimate of the second expectation
Mu1[1] <- 0; Mu2[1] <- 1 ### start values
for(j in 1:10)
{
    for (i in 1:100)
    {
        r[i] <- pi*dnorm(Z[i],Mu2[j],1)
        /((1-pi)*dnorm(Z[i],Mu1[j],1)
        +pi*dnorm(Z[i],Mu2[j],1));
    }
```

```
    Mu1[j+1] <- sum((1-r)*Z)/sum(1-r)
    Mu2[j+1] <- sum(r*Z)/sum(r)
}
```

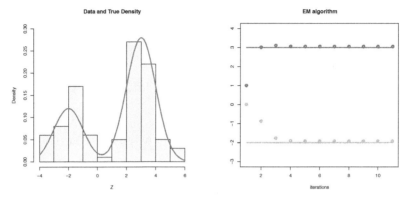

FIGURE 4.3: EM algorithm for normal mixtures.

Note that mixture models deliver good approximation, but the interpretation of the latent variable will be lost.

The EM algorithm becomes a simulation method when the E-step is carried out by a Monte Carlo method. This was proposed in 1990 by Wei and Tanner (1990) as the MCEM algorithm. They assumed that the observed model of $\mathbf{Y}$ has latent variables $\mathbf{Z}$. The complete model is related to $\mathbf{X} = (\mathbf{Y}, \mathbf{Z})$.

Algorithm 4.3 MCEM Algorithm

Given the current state $\theta^{(k)}$.

    1. E-step (Expectation): Calculate $Q(\theta \mid \theta^{(k)})$ by Monte Carlo method:

        (a) Generate independently $\mathbf{z}^{(1)}, \ldots, \mathbf{z}^{(M)}$ from $f(\mathbf{y}, \mathbf{z} \mid \theta^{(k)})$.

        (b) Approximate $Q(\theta \mid \theta^{(k)})$ by

$$Q_{k+1}(\theta \mid \theta^{(k)}) = \frac{1}{M} \sum_{j=1}^{M} \ln(f(\mathbf{y}, \mathbf{z}^{(j)} \mid \theta^{(k)})).$$

    2. M-step (Maximization):

$$\theta^{(k+1)} = \arg\max_{\theta} Q_{k+1}(\theta \mid \theta^{(k)}).$$

4.2 SIMEX

The simulated latent variables $\mathbf{z}^{(1)}, \ldots, \mathbf{z}^{(M)}$ are also called **multiple imputations**. For more details, see Rubin (2004).

4.2 SIMEX

The SIMEX method was introduced in 1994 by Cook and Stefanski (1994). SIMEX stands for Simulation-Extrapolation estimation. It is a measurement error model and a special case of latent variable models. In contrast to the EM algorithm, we observe the latent variable with an additional error (measurement error). SIMEX is a simulation method to get rid of the measurement error in the estimator. The main idea is to increase the measurement error by adding pseudo-errors with stepwise increasing variances. The change of the estimator with respect to the increasing perturbation is modelized. The SIMEX estimator is the backwards extrapolated value to the theoretical point of no measurement error.

> Just make the bad thing worse, study how it behaves and guess the good situation.

Let us introduce the measurement error model and the general principle of SIMEX. Then, we will study the simple linear model with measurement errors in more detail.

Consider a model without measurement errors:

$$(\mathbf{Y}, \mathbf{V}, \mathbf{U}) = (Y_i, V_i, U_i)_{i=1,\ldots,n} \sim F_\theta,$$

with

$$E(Y_i \mid U_i, V_i) = G(\theta, U_i, V_i).$$

Example 4.3 (Blood pressure) Consider a study on the influence of blood pressure on the risk of having a stroke. The data set is given by

$$(Y_i, V_i, X_i)_{i=1,\ldots,n},$$

where Y_i is Bernoulli distributed. The "success" probability is the probability that the ith patient getting a stroke. The variable V_i consists of control variables such as sex, weight, and age of patient i. The variable X_i is the measurement of the blood pressure U_i of patient i. We are interested in estimating the parameter $\theta = (\beta_0, \beta_1^T, \beta_2)$ of

$$E(Y_i \mid U_i, V_i) = \frac{\exp(\beta_0 + \beta_1^T V_i + \beta_2 U_i)}{1 + \exp(\beta_0 + \beta_1^T V_i + \beta_2 U_i)}.$$

$\square$

Another example of a measurement error model is the problem of calibrating two measurement methods.

Example 4.4 (Calibration) Suppose we are interested in comparing two instruments. Each instrument produces another type of measurement error. The aim is to find a relation for transforming data sets from an old instrument and compare them with new data obtained from a new advanced instrument. We have to apply both instruments on the same objects, as the children are doing in Figure 4.4. For each apple i, we observe two weights x_i and y_i. The relationship of interest is between the expected values of the weights:

$$EY_i = \alpha + \beta EX_i.$$

□

FIGURE 4.4: Calibration of a balance with a digital instrument. Example 4.4.

We are interested in estimating θ, and already have a favorite estimator

$$\widehat{\theta} = T(\mathbf{Y}, \mathbf{V}, \mathbf{U}).$$

In the measurement error model, $\mathbf{U}$ is a latent variable, which cannot be directly observed. We observe $\mathbf{X} = (X_1, \ldots, X_n)$ with

$$X_i = U_i + \sigma Z_i, \ \ i = 1, \ldots, n,$$

4.2 SIMEX

where

$$Z_1, \ldots, Z_n \ i.i.d. \text{ are the measurement errors, } Var(Z_i) = \sigma^2.$$

In general, the naive use of the "old" estimation rule

$$\widehat{\theta}_{naive} = T(\mathbf{Y}, \mathbf{V}, \mathbf{X})$$

delivers an inconsistent procedure. SIMEX is a general method for correcting the naive estimator. We assume that σ^2 is known.

Algorithm 4.4 SIMEX

1. Simulation: For every $\lambda \in \{\lambda_1, \ldots, \lambda_K\}$ generate new errors Z_i^* independent on the data and with expectation zero. Calculate new samples

$$X_i(\lambda) = X_i + \sqrt{\lambda}\sigma Z_i^*, \ i = 1, \ldots, n,$$

$$\mathbf{X}(\lambda) = (X_1(\lambda), \ldots, X_n(\lambda)).$$

2. Calculate for every $\lambda \in \{\lambda_1, \ldots, \lambda_K\}$

$$\widehat{\theta}(\lambda) = T(\mathbf{Y}, \mathbf{V}, \mathbf{X}(\lambda)).$$

3. Extrapolation:

Fit a curve $\widehat{f}(\lambda)$ such that $\sum_{k=1}^{K} \left(\widehat{f}(\lambda_k) - \widehat{\theta}(\lambda_k)\right)^2$ is minimal.

4. Define

$$\widehat{\theta}_{simex} = \widehat{f}(-1).$$

Note

$$Var(X_i) = \sigma^2 \text{ and } Var(X_i(\lambda)) = (1+\lambda)\sigma^2$$

such that $\lambda = -1$ respects the case of no measurement error. Obviously, we can only generate new data for positive λ. This is why we need the backwards extrapolation step. The knowledge of the measurement error variance is necessary for this algorithm.

Let us now study the simple linear model with and without measurement errors. Consider the simple linear relationship:

$$y_i = \alpha + \beta \xi_i + \epsilon_i, \quad x_i = \xi_i + \delta_i. \tag{4.10}$$

The $\xi_1, \ldots, \xi_n$ are unknown design points (variables). The first equation is the usual simple linear regression. The second equation is the errors-in-variables

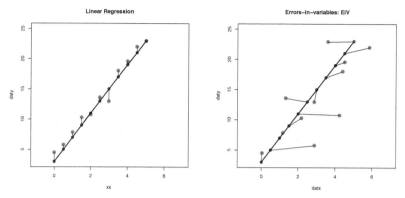

FIGURE 4.5: The simple linear regression model without errors in variables and with errors in variables.

equation. Regression models with observed independent variables are also called errors-in-variables (EIV) models.

In EIV models, the situation is more complicated. The observation (x_i, y_i) can deviate in all directions from the point on the regression line $(\xi_i, \alpha + \beta \xi_i)$. As an example, see Figure 4.5.

The least squares estimators (LSE) in the model with known design points is defined by

$$(\widehat{\alpha}, \widehat{\beta}) = \arg\min_{\alpha,\beta} \sum_{i=1}^{n}(y_i - \alpha - \beta \xi_i)^2.$$

Introduce the usual denotations m_{xx}, m_{xy}, $m_{\xi\xi}$, $m_{\xi y}$ by

$$m_{xx} = \frac{1}{n}\sum_{i=1}^{n}(x_i - \bar{x})^2, \quad m_{xy} = \frac{1}{n}\sum_{i=1}^{n}(x_i - \bar{x})(y_i - \bar{y}),$$

and so on. Then

$$\widehat{\alpha} = \bar{y} - \widehat{\beta}\bar{\xi}, \quad \widehat{\beta} = \frac{m_{y\xi}}{m_{\xi\xi}}.$$

The naive use of this estimator gives

$$\widehat{\alpha}_{naive} = \bar{y} - \widehat{\beta}_{naive}\bar{x}, \quad \widehat{\beta}_{naive} = \frac{m_{xy}}{m_{xx}}.$$

Remind the central limit theorem, we know

$$m_{xy} = \beta m_{\xi\xi} + o_P(1) \text{ and } m_{xx} = m_{\xi\xi} + Var(\delta) + o_P(1).$$

We obtain for the naive estimator

$$\widehat{\beta}_{naive} = \frac{m_{xy}}{m_{xx}} = \frac{\beta m_{\xi\xi}}{m_{\xi\xi} + Var(\delta)} + o_P(1).$$

4.2 SIMEX

In case that the measurement error variance $Var(\delta) = \sigma^2$ is known, we can correct the estimator

$$\widehat{\beta}_{corr} = \frac{m_{xy}}{m_{xx} - \sigma^2}.$$

Assume that the errors (ϵ_i, δ_i) are i.i.d. from a two-dimensional normal distribution with expected values zero, uncorrelated, and $Var(\epsilon) = \sigma^2$ and $Var(\delta) = \kappa\sigma^2$, where κ is known. The maximum likelihood estimator is the total least squares estimator defined by

$$(\widehat{\alpha}_{tls}, \widehat{\beta}_{tls}) = \arg\min_{\alpha,\beta,\xi_1,\ldots,\xi_n} \left(\sum_{i=1}^n (y_i - \alpha - \beta\xi_i)^2 + \kappa \sum_{i=1}^n (x_i - \xi_i)^2 \right).$$

In this simple linear EIV model, the minimization problem can be solved explicitly. The nuisance parameter can be eliminated by orthogonal projection onto the regression line

$$\widehat{\xi}_i = \arg\min_{\xi_i} \left((y_i - \alpha - \beta\xi_i)^2 + \kappa(x_i - \xi_i)^2 \right),$$

that is

$$\widehat{\xi}_i = \frac{\beta y_i + \kappa x_i}{\kappa + \beta^2},$$

see Figure 4.6. Thus, it remains to solve the minimization problem

$$\min_{\alpha,\beta} \frac{1}{\kappa + \beta^2} \sum_{i=1}^n (y_i - \alpha - \beta x_i)^2.$$

We get

$$\widehat{\alpha}_{tls} = \bar{y} - \widehat{\beta}_{tls}\bar{x},$$

and

$$\widehat{\beta}_{tls} = \frac{m_{yy} - \kappa m_{xx} + \sqrt{(m_{yy} - \kappa m_{xx})^2 + 4\kappa m_{xy}^2}}{2m_{xy}}. \tag{4.11}$$

for $m_{xy} \neq 0$, otherwise $\widehat{\beta}_{tls} = 0$.

Algorithm 4.5 SIMEX, Simple Linear EIV

1. Simulation: For every $\lambda \in \{\lambda_1, \ldots, \lambda_K\}$ generate new errors Z_i^* independently on the data and with expectation zero. Calculate new samples

$$X_i(\lambda) = X_i + \sqrt{\lambda}\sigma Z_i^*, \; i = 1, \ldots, n, \quad \mathbf{X}(\lambda) = (X_1(\lambda), \ldots, X_n(\lambda)).$$

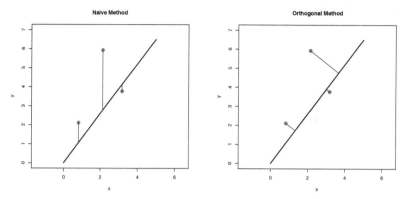

FIGURE 4.6: The unknown design points are estimated by the observation or estimated by the projection on to the line.

> 2. Calculate for every $\lambda \in \{\lambda_1, \ldots, \lambda_K\}$
> $$\widehat{\beta}_{naive}(\lambda) = \frac{m_{x(\lambda)y}}{m_{x(\lambda)x(\lambda)}}.$$
>
> 3. Fit a curve
> $$(\widehat{a}, \widehat{b}) = \arg\min_{a,b} \sum_{k=1}^{K} \left(\frac{a}{b + \lambda_k} - \widehat{\beta}_{naive}(\lambda_k) \right)^2.$$
>
> 4. Define
> $$\widehat{\beta}_{simex} = \frac{\widehat{a}}{\widehat{b} - 1}.$$

It can be shown that
$$\widehat{\beta}_{simex} = \widehat{\beta}_{corr} + o_{P^*}(1).$$
The remainder term $o_{P^*}(1)$ converges to zero in probability, as K tending to infinity with respect to the pseudo-error generating probability P^*.
The model (4.10) is symmetric in both variables, set $EY_i = \eta_i$ then
$$\eta_i = \alpha + \beta \xi_i, \quad \xi_i = -\frac{\alpha}{\beta} + \frac{1}{\beta}\eta_i.$$

The first equation is related to a regression of Y on X, and the second to an inverse regression of X on Y. In case of errors in variables, both regressions are the same. Thus, we have the choice between two different naive estimators
$$\widehat{\beta}_{1,naive} = \frac{m_{xy}}{m_{xx}}, \quad \widehat{\beta}_{2,naive} = \frac{m_{yy}}{m_{xy}}.$$

4.2 SIMEX

It holds $\widehat{\beta}_{1,naive} \leq \widehat{\beta}_{tls} \leq \widehat{\beta}_{2,naive}$.

The following SIMEX algorithm explores this symmetry and is called SYMEX. It has the advantages that the knowledge of the measurement error variance is not needed, but the quotient κ is supposed to be known. The idea is to apply the first SIMEX steps to each of the naive estimators. The SYMEX estimator is defined by the crossing point of the two extrapolation curves.

Algorithm 4.6 SYMEX

1. Regression Y on X

   (a) Simulation: Generate new samples for every $\lambda \in \{\lambda_1, \ldots, \lambda_K\}$

   $$X_i(\lambda) = X_i + \sqrt{\lambda \kappa} Z_{1,i}^*, \ i = 1, \ldots, n,$$

   $$\mathbf{X}(\lambda) = (X_1(\lambda), \ldots, X_n(\lambda)).$$

   (b) Calculate for every $\lambda \in \{\lambda_1, \ldots, \lambda_K\}$

   $$\widehat{\beta}_{1,naive}(\lambda) = \frac{m_{x(\lambda)y}}{m_{x(\lambda)x(\lambda)}}.$$

   (c) Fit a curve $b_1(\lambda) = \frac{\widehat{a}}{b+\lambda}$ with

   $$(\widehat{a}, \widehat{b}) = \arg\min_{a,b} \sum_{k=1}^{K} \left(\frac{a}{b+\lambda_k} - \widehat{\beta}_{naive}(\lambda_k) \right)^2.$$

2. Regression X on Y

   (a) Simulation: Generate new samples for every $\lambda \in \{\lambda_1, \ldots, \lambda_K\}$

   $$Y_i(\lambda) = Y_i + \sqrt{\lambda} Z_{2,i}^*, \ i = 1, \ldots, n,$$

   $$\mathbf{Y}(\lambda) = (Y_1(\lambda), \ldots, Y_n(\lambda)).$$

   (b) Calculate for every $\lambda \in \{\lambda_1, \ldots, \lambda_K\}$

   $$\widehat{\beta}_{2,naive}(\lambda) = \frac{m_{y(\lambda)y(\lambda)}}{m_{y(\lambda)x}}.$$

   (c) Fit a line $b_2(\lambda) = \widehat{a} + \widehat{b}\lambda$ with

   $$(\widehat{a}, \widehat{b}) = \arg\min_{a,b} \sum_{k=1}^{K} \left(b + a\lambda_k - \widehat{\beta}_{naive}(\lambda_k) \right)^2.$$

> 3. Find λ^* such that
> $$b_1(\lambda^*) = b_2(\lambda^*).$$
>
> 4. Define
> $$\widehat{\beta}_{symex} = b_2(\lambda^*).$$

This estimator can be generalized to a multivariate EIV model. For more details, see Polzehl and Zwanzig (2003). It holds

$$\widehat{\beta}_{symex} = \widehat{\beta}_{tls} + o_{P^*}(1),$$

where the remainder term $o_{P^*}(1)$ converges to zero in probability for K tending to infinity, with respect to the pseudo-error generating probability P^*.

The following R code is for a SIMEX algorithm with linear exploration model, see Figure 4.7.

R Code 4.2.24. SIMEX

```
### data=(daty,datx)
### add pseudo-errors with one fixed lamda
### and calculate the naive estimator
simest <- function(n,lamda,B)
{
    b1 <- rep(NA,B)
    for (j in 1:B)
    {
        x <- datx+sqrt(lamda)*rnorm(n,0,1)
        M <- lm(daty~x)
        b1[j] <- coef(M)[2]
    }
    return(b1)
}
simest(11,0.5,10)
mean(simest(11,0.5,10)) ### stabilizing
### SIM - step
l <- c(0.1,0.2,0.3,0.4,0.5,0.6)
K <- length(l)
b <- rep(NA,K)
B <- 1
n <- length(daty)
for(j in 1:K)
{
    b[j] <- mean(simest(n,l[j],B))
}
```

4.3 Variable Selection

```
### EX - step
M1 <- lm(b~l)
### backwards extrapolation, Var=0.5
simb <- coef(M1)[1]-0.5*coef(M1)[2]
```

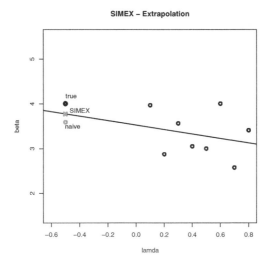

FIGURE 4.7: Illustration of the SIMEX method with a linear extrapolation model.

4.3 Variable Selection

In this section, we consider the problem of variable selection based on simulation. The data set consists of $p+1$ columns of dimension n. Each column includes the observations related to one variable. Y is called the response variable (dependent variable), and the other columns contain the observations of the independent variables (features, covariates) $X_{(1)}, \ldots, X_{(p)}$.

$$\left(\mathbf{Y}, \mathbf{X}_{(1)}, \ldots, \mathbf{X}_{(p)}\right) = \begin{pmatrix} Y_1 & X_{(1),1} & \cdots & X_{(p),1} \\ \vdots & \vdots & \vdots & \vdots \\ Y_n & X_{(1),n} & \cdots & X_{(p),n} \end{pmatrix}. \quad (4.12)$$

The problem is to select the minimal set of covariates $X_{(j_1)}, \ldots, X_{(j_m)}$, needed for predicting the values of Y

$$E\left(Y \mid X_{(1)}, \ldots, X_{(p)}\right) = E\left(Y \mid X_{(j_1)}, \ldots, X_{(j_m)}\right). \quad (4.13)$$

We call the variables $X_{(j_1)}, \ldots, X_{(j_m)}$ in (4.13) *important* and the other variables *unimportant*. The set of indices of the important variables $\mathcal{A} = \{j_1, \ldots, j_m\}$ is called active set.

Example 4.5 (Diabetes) This example is quoted from Efron et al. (2003). 442 diabetes patients were measured on 10 different variables (age, sex, bmi, map, tc, ldl, hdl, tch, ltg, glu) and the response Y is a measurement of disease progression. The problem is to the covariates that are important factors for the disease progression. The results of different methods are illustrated in Figure 4.8, 4.10, 4.11. □

R Code 4.3.25. Diabetes, Example 4.7.

```
library(lars)
data(diabetes)
```

4.3.1 F-Backward and F-Forward Procedures

We briefly review the classical selection methods in linear regression. Let us assume that the data follow a linear regression model

$$\mathbf{Y} = \mathbf{X}\beta + \varepsilon, \ \mathbf{X} = (\mathbf{X}_{(1)}, \ldots, \mathbf{X}_{(p)}), \ rank(\mathbf{X}) = p, \ \varepsilon \sim N_n(0, \sigma^2 \mathbf{I}_n). \quad (4.14)$$

Then the equation (4.13) corresponds to the hypothesis

$$H_0 : \beta_j = 0, \text{ for all } j \notin \mathcal{A}. \quad (4.15)$$

In linear models (4.14) the main tool for testing is the F-test. The F-statistic measures the difference in the model fit of a "small" model with the variables $X_{(j)}, j \in J_m$, and a "big" model with the variables $X_{(j)}, j \in J_q$, where $J_m \subset J_q \subseteq \{1, \ldots, p\}$, $\#(J_m) = m$, $\#(J_q) = q$, $m < q$, and

$$F(J_m, J_q) = \frac{q-m}{n-q} \frac{RSS_{J_m} - RSS_{J_q}}{RSS_{J_q}}.$$

Here the model fit is measured by the residual sum of squares (RSS). Let RSS_J be the residual sum of squares in the linear model including all variables $X_{(j)}, j \in J \subseteq \{1, \ldots, p\}$

$$RSS_J = \min_{\beta, \beta_j = 0, \ j \notin J} \|\mathbf{Y} - \mathbf{X}\beta\|^2.$$

Note, the RSS_J criterion is also reasonable when the linear model assumptions are not fulfilled. Then, for $J = \{j_1, \ldots, j_q\}$ the criterion RSS_J is a measure

4.3 Variable Selection

for the best linear approximation of $E(Y \mid X_{(1)}, \ldots, X_{(p)})$ by a linear function of $X_{(j_1)}, \ldots, X_{(j_q)}$ and

$$RSS_J = \min_{\beta, \beta_j = 0, j \notin J} \left\| E(\mathbf{Y} \mid \mathbf{X}_{(1)}, \ldots, \mathbf{X}_{(p)}) - \mathbf{X}\beta \right\|^2 + tr\left(Cov(Y \mid X_{(1)}, \ldots, X_{(p)})\right) + n\, o_p(1).$$

In (4.14), under the null hypothesis, the small model is true and the F-statistic is F-distributed with $df_1 = n - q$ and $df_2 = q - m$ degrees of freedom. Let $q(\alpha, df_1, df_2)$ be the respective α quantile. Selection algorithms perform combinations of F-tests. In practice, methods which combine backward and forward steps are in use. The backward and forward algorithm will be presented here.

The backward algorithm starts with the model including all the variables of the data set. The first step is a simultaneous test on the p null hypotheses that one component of β in (4.13) is zero. All possible partial F-statistics are compared. One performs the F-test based on the F-statistic with minimal value. No rejection of the null hypothesis means that the respective variable is deleted from the model; otherwise, the whole model is accepted. In the next step, the model with the deleted variable is the big model and all small models with one additional deleted variable are simultaneously tested.

Algorithm 4.7 F-Backward
Choose α.

1. Start with the model (4.14) which includes all variables $X_{(1)}, \ldots, X_{(p)}$.

   (a) For all $j \in J = \{1, \ldots, p\}$ consider $H_{0,j} : \beta_j = 0$ and the partial F-statistic $F_j = F(J \setminus \{j\}, J)$.

   (b) Select l_1 with $F_{l_1} = \min_{j \in J} F_j$.

   (c) Perform the F-test for the hypothesis H_{0, l_1}:

   If $F_{l_1} < q(\alpha, 1, n - p)$, then delete the variable $X_{(l_1)}$ and go to the next step.

   Otherwise stop; select the model with the variables $X_{(1)}, \ldots, X_{(p)}$.

2. Assume the model with the variables $X_{(j)}$, $j \in J_1 = \{1, \ldots, p\} \setminus \{l_1\}$.

   (a) For all $j \in J_1$ calculate the F-statistics $F_j = F(J_1 \setminus \{j\}, J_1)$.

   (b) Select l_2 with $F_{l_2} = \min_{j \in J_1} F_j$.

   (c) If $F_{l_2} < q(\alpha, 1, n - p + 1)$, then delete the variable $X_{(l_2)}$ and go to the next step.

> Otherwise stop; select the model with the variables $X_{(j)}, j \in J_1 = \{1,\ldots,p\} \setminus \{l_1\}$.
>
> 3. Last step: Assume the model with the variable $X_{(l_p)}$, $\{l_p\} = \{1,\ldots,p\} \setminus \{l_1,\ldots,l_{p-1}\}$
>   (a) Consider $H_{0,l_p} : \beta_{l_p} = 0$ and $F_{l_p} = F(\{\},\{l_p\})$
>   (b) If $F_{l_p} < q(\alpha,1,n)$, then delete the variable $X_{(l_p)}$ and no variable is selected.
>
>   Otherwise stop; select the model with the variable $X_{(l_p)}$.

The F-forward selection algorithm is a stepwise regression starting with all possible simple linear regressions. In every step, new models with the new variable are compared by the partial F-statistic. For the model with the maximal F-statistic, the partial F-test at level α is carried out. In the case of rejecting the null hypothesis, the variable with the maximal F-statistic is added to the model. Otherwise, all null hypotheses are not rejected, no more variable is added, and the procedure stops.

Note that forward algorithms have the advantage that they can be applied for big data sets where $p > n$.

> **Algorithm 4.8 F-Forward**
> Choose α.
>
> 1. Suppose p different models with variable $X_{(j)}$, $j \in J = \{1,\ldots,p\}$.
>   (a) In each model calculate the F-statistics $F_j = F(\{\},\{j\})$ for testing $H_{0,j} : \beta_j = 0$.
>   (b) Select j_1 with $F_{j_1} = \max_j F_j$.
>   (c) For $F_{j_1} > q(\alpha,1,n-1)$ select the variable $X_{(j_1)}$ and go to Step 2.
>
>   Otherwise stop, no variable is selected.
>
> 2. Suppose $p-1$ different models each with variables $X_{(j_1)}, X_{(j)}, j \in J_1 = \{1,\ldots,p\} \setminus \{j_1\}$.
>   (a) Calculate the F-statistics $F_{(j_1,j)} = F(\{j_1\},\{j,j_1\})$ for testing $H_{0,j} : \beta_j = 0$.
>   (b) Select j_2 with $F_{(j_1,j_2)} = \max_j F_{(j_1,j)}$.
>   (c) For $F_{j_1,j_2} > q(\alpha,1,n-2)$ select variable $X_{(j_2)}$, and go to Step 3.
>
>   Otherwise stop; the final model contains only $X_{(j_1)}$.

4.3 Variable Selection

> 3. Suppose $p-2$ different models with the variables $X_{(j_1)}, X_{(j_2)}, X_j, j \in J_2 = J_1 \setminus \{j_2\}$.
>
> ...

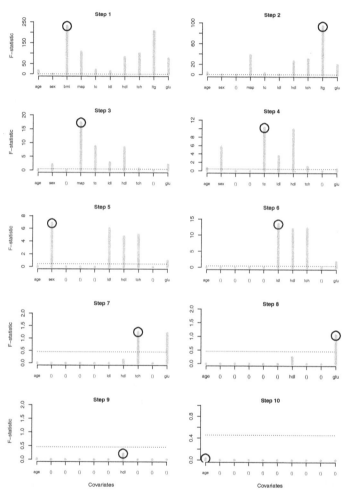

FIGURE 4.8: All steps of the F-forward algorithm in Example 4.5. The broken line is the $\alpha = 0.5$ quantile. The algorithm stops after Step 8. The selected variables are bmi, ltg, map, tc, sex, ldl, tch, glu.

The tuning parameter α is the size of every partial F-test. The final model is based on several tests, and α is not anymore the size of the whole procedure. The size of the whole procedure is called the *family size* and defined by

$$\alpha_{family} = P(\text{the true model is not selected}).$$

As an example, see Figure 4.9 for different values of α. In case of the backward algorithm, a bound of the real family size can be calculated. Suppose the chosen model contains q variables, then the decisions $D_1, \ldots, D_{p-q+1}$ are made, where only the last decision is a rejection of the partial null hypothesis. Using

$$P_0(D_1, \ldots, D_{p-q+1})$$
$$= P_0(D_{p-q+1} \mid D_1, \ldots, D_{p-q}) P_0(D_{p-q} \mid D_1, \ldots, D_{p-q-1}) \cdots P_0(D_1),$$

we have

$$P(\text{model is selected} \mid \text{model is true})$$
$$= P_0\left(\beta_j = 0, j \in J = \mathcal{A} \text{ is not rejected}, \beta_{l_{q+1}} = 0 \text{ is rejected}\right)$$
$$= (1-\alpha)^{p-q}\alpha.$$

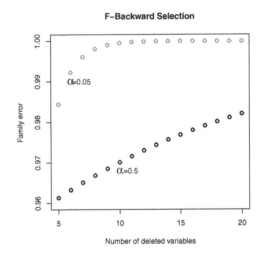

FIGURE 4.9: F-backward selection for different values of α.

The main problem is now to determine the tuning parameter α which controls the threshold of the model fit criterion. The F-statistic, as a measure of the influence of covariates, is also useful in case of nonlinear models. Then it compares the best linear approximations. If the condition on the error distribution is violated, the F-statistics are not F-distributed. The comparison of the F-statistic with a bound $q(\alpha, df_1, df_2)$ controlled by the tuning parameter α is still reasonable. The bound is now simply a threshold and not anymore the quantile of the F distribution F_{df_1, df_2}. The tuning parameter α is not the size of the partial test.

4.3 Variable Selection

In R, variable selection procedures are implemented where the models are compared by Akaike information criterion (AIC) or the Bayesian information criterion (BIC). For a linear model with normal errors, unknown error variance and included variables $X_{(j)}, j \in J = \{j_1, \ldots, j_q\}$, the AIC criterion is defined as

$$AIC_J = n\ln(RSS_J) + 2q,$$

and the BIC

$$BIC_J = n\ln(RSS_J) + \ln n\, q.$$

Note that in some cases the criteria differ by an additive constant. An alternative is Mallows C_p

$$C_{p,J} = \frac{RSS_J}{RSS_{\{1,\ldots,p\}}} \frac{n-p}{1} + 2q.$$

R Code 4.3.26. Variable selection, Example 4.5.

```
library(lars)
data(diabetes)
library(leaps) ### subset selection
all <- regsubsets(y~x,data=diabetes)
### stepwise adding variables can be done as follows
### step1
lm0 <- lm(y~1)
A1 <- add1(lm0,~1+x[,1]+x[,2]+x[,3]+x[,4]+x[,5]+
           x[,6]+x[,7]+x[,8]+x[,9]+x[,10],test="F")
F1 <- A1$F[2:11] ### partial F statistic for every new added
    variable
AIC.1 <- A1$AIC[2:11] ### AIC for every new added variable
A1$AIC[1] ### minimal AIC of the step before
max(F1)
which.max(F1) ### here at bmi= x[,3]
### step2
lm1 <- lm(y~1+x[,3],data=diabetes)
A2 <- add1(lm1,~x[,1]+x[,2]+x[,3]+x[,4]+x[,5]+
           x[,6]+x[,7]+x[,8]+x[,9]+x[,10],test="F")
```

The AIC-forward algorithm is a stepwise minimization algorithm of the AIC criterion. The model with the smallest AIC is taken for the next step. The procedure stops when no smaller AIC can be reached.

> **Algorithm 4.9 AIC-Forward**
>
> 1. Let $J_q = \{j_1, \ldots, j_q\}$ be the current set of selected variables and $AIC_{(q)} = AIC_{J_q}$.
> 2. For all $j \in \{1, \ldots, p\} \setminus J_q$ calculate the AIC-statistics
>
> $$AIC_{J_q \cup j} = n\ln(RSS_{J_q \cup j}) + 2n(q+1).$$
>
> 3. Select j_{q+1} with
>
> $$AIC_{J_q \cup j_{q+1}} = \min_j AIC_{J_q \cup j},$$
>
> and set $AIC_{(q+1)} = AIC_{J_q \cup j_{q+1}}$.
>
> 4. For $AIC_{(q+1)} < AIC_{(q)}$ select the variable $X_{(j_{q+1})}$ and update $J_{q+1} = J_q \cup \{j_{q+1}\}$, otherwise stop.

Both forward algorithms are greedy algorithms, because in every step the next step is optimized. In every step, we compare models with the same number of variables and

$$RSS_1 < RSS_2 \Leftrightarrow AIC_1 < AIC_2$$
$$\Leftrightarrow BIC_1 < BIC_2$$
$$\Leftrightarrow F_2 < F_1.$$

Thus, there is no difference in selecting the next variable between F-forward, AIC-forward or BIC-forward algorithm. The procedures differ in their stopping rules. In case that the number q_{true} of variables in the true model is known, and it is only unknown which of the p variables in the data set should be included, all methods stop after selecting q_{true} variables and deliver the same final model. The assumption that the active set $\mathcal{A}$ contains not more than q_{true} elements is called sparsity assumption.

Note that the equivalences work for all criteria, which are monotone functions of RSS, especially for criteria which penalize the RSS by an additive constant. The equivalence does not hold for a criterion of the following type

$$\min_{\beta, \beta_j = 0, j \notin J} (\|\mathbf{Y} - \mathbf{X}\beta\|^2 + \lambda pen(\beta, q)),$$

where the penalty depends on the unknown parameter β.

4.3.2 FSR-Forward Procedure

The next procedure applies data augmenting for determining the stopping condition. In Miller (2002), he proposed the generation of pseudo-variables for

4.3 Variable Selection

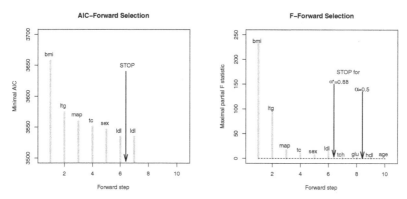

FIGURE 4.10: Example 4.5. The AIC-forward algorithm stops after the 6th step. The selected variables are bmi, ltg, map, tc, sex, ldl. Using the FSR method with α^*, the selected variables are bmi, ltg, map, tc, sex, ldl. In this example, both methods deliver the same final model.

calibrating forward selection procedures. Instead of the original data (4.12), extended data sets are introduced. Let

$$\left(\mathbf{Y}, \mathbf{X}_{(1)}, ..., \mathbf{X}_{(p)}, \mathbf{Z}_{(1)}, ..., \mathbf{Z}_{(K)}\right) =$$

$$\begin{pmatrix} Y_1 & X_{(1),1} & \cdots & X_{(p),1} & Z_{(1),1} & \cdots & Z_{(K),1} \\ \vdots & \vdots & \vdots & \vdots & \vdots & \vdots & \vdots \\ Y_n & X_{(1),n} & \cdots & X_{(p),1} & Z_{(1),n} & \cdots & Z_{(K),n} \end{pmatrix},$$

where the observations $Z_{(j),i}, j = 1, \ldots, K$, $i = 1, \ldots, n$ are generated such that the generated variables $Z_{(1)}, \ldots, Z_{(K)}$ are unimportant by construction. For example, generate $Z_{(j),i}$ i.i.d. standard normal distributed. In the literature, they are called phony variables, control variables, or pseudo-variables. The arguments are as follows.

A procedure which selects many phony variables will select many unimportant variables. Otherwise, a procedure which never accepts a phony variable has the big risk to miss an important variable.

Wu et al. (2007) developed an algorithm in his dissertation and called it FSR-algorithm. FSR stands for false selecting rate. Running a selection procedure on the data $(\mathbf{Y}, \mathbf{X})$, let $U(\mathbf{Y}, \mathbf{X})$ be the number of unimportant selected variables and $S(\mathbf{Y}, \mathbf{X})$ the number of all selected variables, then

$$FSR = \frac{U(\mathbf{Y}, \mathbf{X})}{S(\mathbf{Y}, \mathbf{X}) + 1}.$$

It is not possible to count $U(\mathbf{Y}, \mathbf{X})$, but the number of selected phony variables is known. Wu's main idea is to determine the size α of the partial F-tests in the F-forward procedure by checking how many of the phony variables are selected, see Figure 4.12.

Algorithm 4.10 FSR-Forward
Given a bound γ_0 on the rate of selected phone variables.

1. For every $\alpha \in \{\alpha_1, \ldots, \alpha_M\}$.

   (a) For every b, $b = 1, \ldots, B$.

      i. Generate $\mathbf{Z}_b = (\mathbf{Z}_{(1)}, \ldots, \mathbf{Z}_{(K)})$ independent of $(\mathbf{Y}, \mathbf{X})$.
      ii. Run a F-forward selection with α on the data $(\mathbf{Y}, \mathbf{X}, \mathbf{Z}_b)$
      and count $U_b^*(\alpha) = \#\{k : \mathbf{Z}_{(k)}\text{ selected}\}$ and count $S_b(\alpha)$ the number of all selected variables.

   (b) Calculate the averages

$$\overline{U}^*(\alpha) = \frac{1}{B}\sum_{j=1}^{B} U_b^*(\alpha), \quad \overline{S}(\alpha) = \frac{1}{B}\sum_{j=1}^{B} S_b(\alpha),$$

   and as an indicator for the FSR

$$\gamma(\alpha) = \frac{\overline{U}^*(\alpha)}{\overline{S}(\alpha) + 1}.$$

2. Determine the adjusted α level

$$\alpha^* = \sup_{\alpha \in \{\alpha_1, \ldots, \alpha_M\}} \{\alpha : \gamma(\alpha) \leq \gamma_0\}.$$

3. Run an F-forward selection on the original data $(\mathbf{Y}, \mathbf{X})$ with α^*.

4.3.3 SimSel

SimSel applies both the data augmentation of Wu's FSR algorithm and the successive data perturbation of SIMEX, introduced in Eklund, Zwanzig (2011). The main idea behind SimSel is to determine the relevance of a variable $\mathbf{X}_{(j)}$ by successively disturbing it, and study the effect on the residual sum of squares (RSS). In case that the RSS remains unchanged, and we conclude that the variable $\mathbf{X}_{(j)}$ is not important.

The SimSel algorithm borrows the simulation step where pseudo-errors are added to the independent variables from the SIMEX method. However, the extrapolation step in SIMEX is not performed. Thus, it is called SimSel for

4.3 Variable Selection

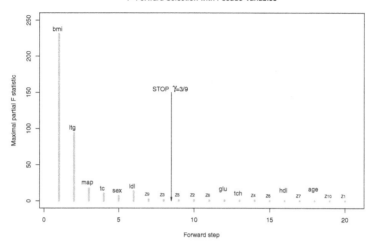

FIGURE 4.11: All possible steps of an F-Forward algorithm with 10 pseudo-variables. Example 4.5.

simulation and selection. The variables $\mathbf{X}_{(1)}, \ldots, \mathbf{X}_{(p)}$ are checked each after the other. Let us assume that we are interested in $\mathbf{X}_{(1)}$. The original data set

$$\left(\mathbf{Y}, \mathbf{X}_{(1)}, \ldots, \mathbf{X}_{(p)}\right)$$

is embedded into

$$\left(\mathbf{Y}, \mathbf{X}_{(1)} + \sqrt{\lambda}\varepsilon^*, \ldots, \mathbf{X}_{(p)}, \mathbf{Z}\right),$$

where $\mathbf{Z} = (Z_1, \ldots, Z_n)^T$ is a **pseudo-variable**, independently generated from $\mathbf{Y}, \mathbf{X}_1, \ldots, \mathbf{X}_p$, the pseudo-errors are generated such that $\varepsilon^* = (\varepsilon_1^*, \ldots, \varepsilon_n^*)^T$, ε_i^* are i.i.d. P^*, with $E(\varepsilon_i^*) = 0$, $Var(\varepsilon_i^*) = 1$, $E(\varepsilon_i^*)^4 = \mu$. The phony variable $\mathbf{Z}$ serves as an untreated control group in a biological experiment. The influence of the pseudo-errors is controlled by stepwise increasing λ. The main idea is, if λ "does not matter", then $\mathbf{X}_{(1)}$ is unimportant, see Figure 4.13. There, we consider the simple linear case $\mathbf{Y} = \beta_1 \mathbf{X}_1 + \varepsilon$, and we fit

$$RSS_1(\lambda_k) = \min_{\beta_1, \beta_2} \left\| \mathbf{Y} - \beta_1 \left(\mathbf{X}_1 + \sqrt{\lambda_k}\varepsilon^*\right) - \beta_2 \mathbf{Z} \right\|^2,$$

and

$$RSS_2(\lambda_k) = \min_{\beta_1, \beta_2} \left\| \mathbf{Y} - \beta_1 \mathbf{X}_1 - \beta_2 \left(\mathbf{Z} + \sqrt{\lambda_k}\varepsilon^*\right) \right\|^2$$

by a linear regression. Intuitively "does not matter" respects to a constant trend of $RSS(.)$.

The theoretical background to the SimSel algorithm is given by the following result:

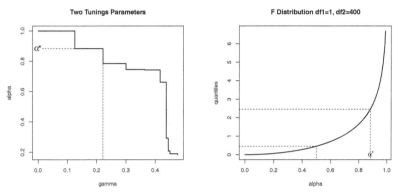

FIGURE 4.12: Determining the α^* of the F-forward algorithm by the FSR estimate from the extended data set.

Theorem 4.3 *Under the assumption that $(\mathbf{X}^T\mathbf{X})^{-1}$ exists, it holds*

$$\frac{1}{n}RSS(\lambda) = \frac{1}{n}RSS + \frac{\lambda}{1+h_{11}\lambda}\left(\widehat{\beta}_1\right)^2 + o_{P^*}(1),$$

where h_{11} is the $(1,1)$–element of $\left(\frac{1}{n}\mathbf{X}^T\mathbf{X}\right)^{-1}$ and $\widehat{\beta}_1$ is the first component of the LSE estimator $\widehat{\beta} = (\mathbf{X}^T\mathbf{X})^{-1}\mathbf{X}^T\mathbf{Y}$.

Proof: It holds

$$\frac{1}{n}RSS(\lambda) = \frac{1}{n}\mathbf{Y}^T\mathbf{Y} - \frac{1}{n}\mathbf{y}^T P(\lambda)\mathbf{Y}, \qquad (4.16)$$

with

$$P(\lambda) = \mathbf{X}(\lambda)\left(\mathbf{X}(\lambda)^T\mathbf{X}(\lambda)\right)^{-1}\mathbf{X}(\lambda)^T. \qquad (4.17)$$

Further

$$\frac{1}{n}\mathbf{X}(\lambda)^T\mathbf{Y} = \left(\frac{1}{n}\mathbf{X} + \frac{1}{n}\sqrt{\lambda}\Delta\right)^T\mathbf{Y},$$

where Δ is the $(n \times p)-$ matrix

$$\Delta = \begin{pmatrix} \varepsilon_1^* & 0 & \cdots & 0 \\ \varepsilon_2^* & 0 & \cdots & 0 \\ \vdots & \vdots & 0 & \vdots \\ \varepsilon_{n-1}^* & 0 & \cdots & \vdots \\ \varepsilon_n^* & 0 & \cdots & 0 \end{pmatrix}.$$

The law of large number applied to the pseudo-errors delivers

$$\frac{1}{n}\mathbf{X}(\lambda)^T\mathbf{Y} = \frac{1}{n}\mathbf{X}^T\mathbf{Y} + o_{P^*}(1). \qquad (4.18)$$

4.3 Variable Selection

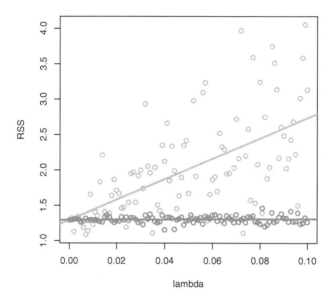

FIGURE 4.13: The constant regression respects to the unimportance of the pseudo-variable. The heteroscedastic regression is related to the worse model fit under the disturbed important variable.

Consider now $\mathbf{X}(\lambda)^T\mathbf{X}(\lambda)$:

$$\frac{1}{n}\left(\mathbf{X}+\sqrt{\lambda}\Delta\right)^T\left(\mathbf{X}+\sqrt{\lambda}\Delta\right) \qquad (4.19)$$

$$= \frac{1}{n}\mathbf{X}^T\mathbf{X} + \frac{1}{n}\sqrt{\lambda}\mathbf{X}^T\Delta + \frac{1}{n}\sqrt{\lambda}\Delta^T\mathbf{X} + \frac{1}{n}\lambda\Delta^T\Delta. \qquad (4.20)$$

Hence

$$\left(\frac{1}{n}\mathbf{X}(\lambda)^T\mathbf{X}(\lambda)\right)^{-1} = \left(\frac{1}{n}\mathbf{X}^T\mathbf{X} + \lambda\mathbf{e}_1\mathbf{e}_1^T\right)^{-1} + o_{P^*}(1),$$

where e_1 is a vector with the first element 1, and the rest 0. □

Note, in the procedure we approximate

$$\frac{\lambda}{1+h_{11}\lambda} \approx \lambda.$$

In linear errors-in-variable models, the naive LSE is inconsistent. But if β_1 is zero, then naive LSE also converges to zero. This gives the motivation for a successful application of SimSel to EIV models.

The significance of the perturbation on the model fit is controlled by a Monte Carlo F-test for the regression of $RSS(\lambda)$ on λ. That is the testing step in the SimSel procedure. The comparison of the model fit is repeated M times and a paired sample of F- statistics are obtained, where sample $F_{i,1}, \ldots, F_{i,M}$ related to the variable under control $\mathbf{X}_{(i)}$. The other samples $F_{p+1,1}, \ldots, F_{p+1,M}$ are related to the pseudo-variable $\mathbf{Z} = \mathbf{X}_{(p+1)}$. Kernel estimates $\widehat{f}_i$, $\widehat{f}_{p+1}$ for each sample are calculated and the overlapping of the densities are compared with given tuning parameters α_1, α_2, see Figure 4.14.

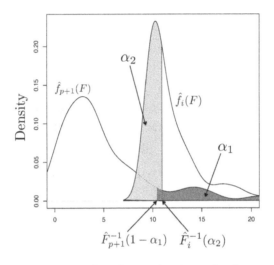

FIGURE 4.14: Significance for chosen levels α_1, α_2.

Example 4.6 (Prostate) The prostate cancer data set contains 97 observations of one dependent variable (the log of the level of prostate-specific antigen, lpsa) and eight independent variables; the logarithm of the cancer volume (lcavol), the logarithm of the prostate's weight (lweight), age, the logarithm of the amount of benign prostatic hyperplasia (lbph), seminal vesicle invasion (svi), the logarithm of the level of capsular penetration (lcp), Gleason score (gleason), and percentage Gleason scores 4 and 5 (pgg45). For more details, see the R package ElemStatLearn, data(prostate). The results of SimSel for this data set are given in Figure 4.15. □

4.3 Variable Selection

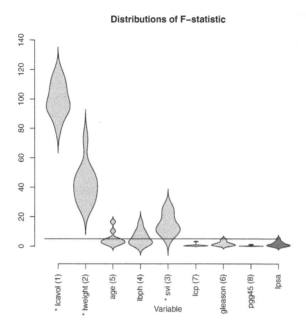

FIGURE 4.15: Graphical output of SimSel. Example 4.6. The important variables are lcavol, lweight, svi.

Algorithm 4.11 The SimSel Algorithm
Given $0 \leq \lambda_1 \leq \lambda_2 \leq ... \leq \lambda_K, M, \alpha_1, \alpha_2$.
For m from 1 to M:
  Generate unimportant pseudo-variables $\mathbf{Z} = \mathbf{X}_{p+1}$.
  For i from 1 to $p+1$:
    For k from 1 to K:
      Generate pseudo-errors for each $\mathbf{Z} = \mathbf{X}_{p+1}$ to $\mathbf{X}_i$ and add them.
      Compute $RSS_i(\lambda_k)$.
    Regression step. Calculate the F-statistics $F_{i,m}$.
Plotting step, violin plot of the F-statistics.
Ranking step, according to the sample median of $(F_{i,m})_{m=1...M}$.
Testing step.

4.4 Problems

1. EM Algorithm: Consider the Hardy-Weinberg model with 6 genotypes. Let $\theta_1, \theta_2, \theta_3$ denote the probabilities of the alleles S, I, F with $\theta_1 + \theta_2 + \theta_3 = 1$. The Hardy-Weinberg model specifies that the six genotypes have the probabilities:

$$\begin{array}{lcccccc} \text{Genotype} & SS & II & FF & SI & SF & IF \\ \text{Probability} & \theta_1^2 & \theta_2^2 & \theta_3^2 & 2\theta_1\theta_2 & 2\theta_1\theta_3 & 2\theta_2\theta_3 \end{array}. \quad (4.21)$$

The data are observed from the following model:

$$\begin{array}{lcccc} \text{Genotype} & SS+II+FF & SI & SF & IF \\ \text{Probability} & \theta_1^2+\theta_2^2+\theta_3^2 & 2\theta_1\theta_2 & 2\theta_1\theta_3 & 2\theta_2\theta_3 \end{array}. \quad (4.22)$$

(a) Calculate the maximum likelihood estimators for $\theta_1, \theta_2, \theta_3$ in the model (4.21).

(b) Write an R code for the EM algorithm when the data are coming from (4.22).

(c) Use the EM algorithm for the data set:

$$\begin{array}{lcccc} \text{Genotype} & SS+II+FF & SI & SF & IF \\ \text{observed} & 164 & 41 & 35 & 160 \end{array}. \quad (4.23)$$

2. Snake's head (Kungsängslilja, Fritillaria meleagris) is selected to represent the province Uppland in Sweden. The flowers were discovered in Uppsala in 1743. Each year in May, the Uppsala surrounding is covered in Snake's head, in three different colors: violet, white, and rose.

Suppose a botanist is interested in the distribution of the colors. By mistake, the flowers of white and rose colors are calculated together.

Suppose that the colors are distributed as follows

$$\begin{array}{ccc} white & rose & violet \\ 1-4p & p & 3p \end{array}.$$

(a) Formulate the complete model for the numbers n_1, n_2, n_3 of flowers of the respective color.

(b) Formulate the model for the simplified observations.

(c) Describe the expectation step of the EM Algorithm in this case.

(d) Describe the maximization step of the EM Algorithm in this case.

4.4 Problems

3. Consider the data $x_i = (A_i, B_i)$ from the **spatial data set:** Twenty-six neurologically impaired children have taken two sets of spatial perception, called "A" and "B".

i	1	2	3	4	5	6	7	8	9	10	11	12	13
A	48	36	20	29	42	42	20	42	22	41	45	14	6
B	42	33	16	39	38	36	15	33	20	43	34	22	7

i	14	15	16	17	18	19	20	21	22	23	24	25	26
A	0	33	28	34	4	32	24	47	41	24	26	30	41
B	15	34	29	41	13	38	25	27	41	28	14	28	40

   We are interested in the relation between A and B in the spatial data set. We assume an errors-in-variable (EIV) model:

   $$A_i = \alpha_i + \varepsilon_i; \ B_i = \beta_i + \delta_i,$$

   and

   $$\alpha_i = a + b\beta_i. \tag{4.24}$$

   The aim is to estimate b.

   (a) Plot the data. Fit a line using the naive least squares and plot it.
   (b) Apply the SIMEX procedure for correcting the naive estimator. Plot the line and compare it with part (a).
   (c) Propose several combined Simex/Bootstrap algorithms for constructing a basic bootstrap interval for the slope b in the EIV model (4.24).
   (d) Calculate a confidence interval for b.
   (e) Is $b = 1$ included?

4. Let $X_1, \ldots, X_n$, i.i.d., be realizations of a random variable X with density

   $$f(x) = \alpha_1 \varphi_{(\mu_1, 1)}(x) + \alpha_2 \varphi_{(3\mu_2, 1)}(x) + (1 - \alpha_1 - \alpha_2)\varphi_{(3\mu_1, 1)}(x),$$

   where $\varphi_{(\mu, 1)}(x)$ is the density of $N(\mu, 1)$, $0 < \alpha_1 < 1, 0 < \alpha_2 < 1$ and $\alpha_1 + \alpha_2 < 1$. We are interested in an EM-estimator of (μ_1, μ_2).

   (a) Formulate a latent variable model. Which distribution has the latent variable?
   (b) Calculate the surrogate function. Define the responsibilities.
   (c) Calculate an estimator of (μ_1, μ_2).
   (d) Write the main steps of the EM algorithm using the results in (b) and (c).

5

Density Estimation

In this chapter, we consider the problem of estimating density. In previous chapters, we used the violin plot as a graphical method for comparing different simulation methods. The violin plot is based on the kernel density estimation. The main problem is that we have weak model assumptions. We have an i.i.d. sample of n observation from a distribution with continuous density, and we want to estimate an almost arbitrary function with the help of these n observation points.

5.1 Background

In classical statistical inference, when a parametric model is proposed, we require that the unknown underlying distribution of the observations belongs to a parametrized family. For instance, when we suppose $X_1, \ldots, X_n$ are i.i.d. from $X \sim N(\mu, \sigma^2)$, then estimating the density is equal to the problem of estimating μ and σ^2. We know that the sample mean $\bar{X}$ and the sample variance s^2 are the best unbiased estimators of μ and σ^2. See for instance Liero and Zwanzig (2011).

To estimate density, we use now a nonparametric setup. Let $\mathbf{X} = (X_1, \ldots, X_n)$ be an i.i.d. sample of the r.v. X with an unknown density

$$f \in \mathcal{F} = \left\{ f : \mathbb{R} \to \mathbb{R}, \ f(x) \geq 0, \int f(x)dx = 1, \ f \text{ continuous} \right\}.$$

We are looking for a function $\widehat{f}$ in $\mathcal{G}$, where

$$\begin{aligned} \mathcal{G} &= \{\widehat{f} : \mathbb{R} \to \mathbb{R}, \ \widehat{f}(x) = \widehat{f}(x, X_1, \ldots, X_n), \\ &\quad \text{continuous in } x \text{ and measurable in } X_1, \ldots, X_n, \\ &\quad \widehat{f}(x) \geq 0, \int \widehat{f}(x)dx = 1\}. \end{aligned}$$

Similar to a classical parametric inference, we consider the mean squared error (MSE) as a criterion for the goodness of estimation of f at an arbitrary given point x

$$MSE\left(\widehat{f}(x)\right) = E\left(\widehat{f}(x) - f(x)\right)^2 = Var(\widehat{f}(x)) + Bias(\widehat{f}(x))^2,$$

with
$$Bias(\widehat{f}(x)) = E(\widehat{f}(x)) - f(x).$$

We can use the mean integrated squared error (MISE) as criteria for a good estimation of f at all points

$$MISE\left(\widehat{f}\right) = E \int \left(\widehat{f}(x) - f(x)\right)^2 dx.$$

Following the line of estimation theory for parametric models, we are interested in an unbiased estimation of $f(x)$ for all x in order to avoid the bias term in MISE. The following theorem indicates that this requirement is unrealistic, see Rosenblatt (1956).

> **Theorem 5.1 (Rosenblatt (1956))** *Assume* $\mathbf{X} = (X_1, \ldots, X_n)$ *i.i.d. sample of the r.v. X with unknown density f. Then there exists no $\widehat{f} \in \mathcal{G}$ such that*
> $$E\widehat{f}(x) = f(x), \text{ for all } x \in \mathbb{R}, \text{ and for all } f \in \mathcal{F}.$$

Proof: We present an indirect proof. Let $\tilde{f} \in \mathcal{G}$ be such that

$$E\tilde{f}(x) = f(x), \text{ for all } x \in \mathbb{R}, \text{ for all } f \in \mathcal{F}.$$

Then, for an arbitrary interval $[a, b]$

$$\tilde{P}[a, b] = \int_a^b \tilde{f}(x)dx$$

is an unbiased estimator of $P[a, b]$ with

$$E(\tilde{P}[a, b]) = P[a, b] = F(b) - F(a),$$

where F is the distribution function. Otherwise, the empirical distribution function $\widehat{F}_n$ is an unbiased estimator of F, such that

$$\widehat{P}[a, b] = \widehat{F}_n(b) - \widehat{F}_n(a)$$

is also an unbiased estimator of $P[a, b]$. The estimator $\widehat{P}$ depends on the order statistic $\mathbf{X}_{[\,]} = (X_{[1]}, \ldots, X_{[n]})$ which is sufficient and complete for the model setup above. For more details, see Bell et al. (1960). This implies that the unbiased estimator is unique. Hence, $\tilde{P}[a, b] = \widehat{P}[a, b]$ and

$$\widehat{F}_n(b) - \widehat{F}_n(a) = \int_a^b \tilde{f}(x)dx.$$

But the empirical distribution function is a step function, and there is no continuous function $\tilde{f}$ that fulfills this equation. That gives the contradiction. □

5.2 Histogram

The underlying reason for this statement is that the requirement of unbiasedness is very strong for big models. The nonparametric model $\mathcal{F}$ is definitely big.

Based on this fact, we apply two principles for estimating a function:

- Approximation

- Estimation of the approximated function

Consider the estimation of $f \in \mathcal{F}$ at an arbitrary given value x. First, we choose a local neighborhood $[a, b]$, with $x \in [a, b]$, and apply the mean value theorem of integration theory

$$f(x) \approx \frac{\int_a^b f(u) du}{b - a} = \frac{P([a,b])}{b - a}.$$

Then, we estimate $P([a, b])$ for instance by

$$\widehat{P}[a, b] = \hat{F}_n(b) - \hat{F}_n(a) = \frac{\#\{i : X_i \in [a, b]\}}{n},$$

where

$$\#\{i : X_i \in [a, b]\} \sim Bin(n, p), \quad p = P([a, b]).$$

The estimate $\widehat{P}[a, b]$ has a small variance for large p, which is the case when the interval $[a, b]$ is large, but a large local neighborhood $[a, b]$ implies a high approximation error. Both principles for approximation and estimation behave contradictorily. When the estimator behaves badly and has a large variance, this is called undersmoothing; the opposite case where the bias becomes too large is called oversmoothing. The trick is to find a balance.

5.2 Histogram

The standard method for estimating density is with a histogram. Here the real line is divided into bins with binwidth $h > 0$. Let x_0 be the origin of the histogram. We define the partition by

$$B_j = [x_0 + (j - 1)h, x_0 + jh], \quad j \in \mathbb{Z}.$$

This means that the distribution F with density f is approximated by a discrete distribution with probability function $p(j) = P(B_j)$. The discrete distribution is estimated by $\frac{1}{n} \sum_{i=1}^{n} I(X_i \in B_j)$, where

$$I(x \in B_j) = \begin{cases} 1 & \text{for } x \in B_j \\ 0 & \text{for } x \notin B_j \end{cases}.$$

144   5 Density Estimation

The histogram is given by

$$\widehat{f}_h(x) = \frac{1}{nh} \sum_{i=1}^{n} \sum_{j} I(X_i \in B_j) I(x \in B_j).$$

Note that the histogram can also be defined with the help of other partitions of the real line. The main point is that the partition is chosen independently from x. The main disadvantage of the histogram is that the continuous distribution is estimated by a discrete distribution. We have $\widehat{f}_h(x) \notin \mathcal{G}$. For more details see the following R code and Figures 5.1, 5.2, and 5.3.

R Code 5.2.27. Histogram

```
z <- rbinom(200,1,0.3)
x <- z*rnorm(200,2,0.5)+(1-z)*rnorm(200,4,0.4) ### simulate data
hist(x,col=grey(0.9)) ### histogram default
hist(x,col=grey(0.9),nclass=16) ### define number of classes
xx <- sort(x)
breaks <- c(xx[1],xx[10],xx[20],xx[30],xx[40],xx[50],xx[60]
    ,xx[80],xx[100],xx[120],xx[140],xx[160],xx[180],xx[200])
hist(x,col=grey(0.9),breaks=breaks) ###  adaptive classes
```

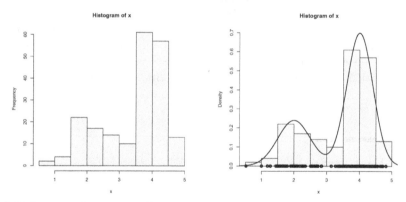

FIGURE 5.1: Left: Histogram with default options. Right: Histogram and true underlying density and data points.

For smooth densities, we apply a Taylor approximation. Then the bias and the variance of the histogram can be approximated by

$$Bias(\widehat{f}_h(x)) = (b(x) - x)) f'(b(x)) + o(h), \quad h \to 0,$$

$$Var(\widehat{f}_h(x)) = \frac{1}{nh} f(x) + o(\frac{1}{nh}), \quad nh \to \infty$$

5.3 Kernel Density Estimator

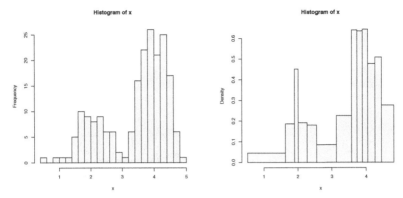

FIGURE 5.2: Left: Histogram with a determined number of classes. Right: Histogram with adaptive chosen classes.

for $x \in B_{j(x)}$, where $b(x) = (j(x) - \frac{1}{2})h$ denotes the midpoint of the bin $B_{j(x)}$. For the mean squared error, we get

$$MSE\left(\widehat{f}_h(x)\right) = \frac{1}{nh}f(x) + (b(x) - x)^2 \left(f'(b(x))\right)^2 + o(\frac{1}{nh}) + o(h^2).$$

For the mean integrated squared error, we get

$$MISE\left(\widehat{f}_h\right) = \frac{1}{nh} + \frac{h^2}{12}\left\|f'\right\|^2 + o(\frac{1}{nh}) + o(h^2), \tag{5.1}$$

where $\left\|f'\right\|^2 = \int f'(x)^2 dx$. The leading term in (5.1) is usually defined as asymptotic MISE (AMISE):

$$AMISE\left(\widehat{f}_h\right) = \frac{1}{nh} + \frac{h^2}{12}\left\|f'\right\|^2.$$

Considering $AMISE\left(\widehat{f}_h\right) = A(h)$ as a function of h, then

$$h_0 = \arg\min A(h) = \left(\frac{6}{n\left\|f'\right\|^2}\right)^{\frac{1}{3}},$$

and $MISE\left(\widehat{f}_{h_0}\right) \approx n^{-\frac{2}{3}}$.

5.3 Kernel Density Estimator

The first approach for improving the histogram for $f(x)$ is to choose a local neighborhood $B(x) = [x - \frac{1}{2}h, x + \frac{1}{2}h]$ around x. This means that for every

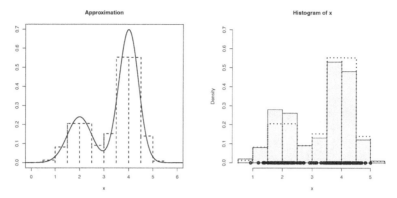

FIGURE 5.3: Left: Approximation of the continuous density by a probability function. Right: Estimation of the dotted approximation.

x, a local approximation is applied, see Figure 5.4. We apply

$$f(x) \approx \frac{\int_{x-\frac{1}{2}h}^{x+\frac{1}{2}h} f(u)du}{h} = \int f(u)\frac{1}{h}U(\frac{x-u}{h})du.$$

where $U(x)$ is the density of the uniform distribution over $[-\frac{1}{2}, \frac{1}{2}]$. Then $\frac{1}{h}U(\frac{x-u}{h})$ is the density of the uniform distribution over $[x - \frac{1}{2}h, x + \frac{1}{2}h]$. We have

$$\int f(u)\frac{1}{h}U(\frac{x-u}{h})du = E_f \frac{1}{h}U(\frac{x-X}{h}),$$

and the local approximation can be estimated by the average

$$\frac{1}{n}\sum_{i=1}^{n}\frac{1}{h}U(\frac{x-X_i}{h}).$$

Thus, the estimator is given by

$$\widehat{f}(x) = \frac{1}{nh}\sum_{i=1}^{n}U(\frac{x-X_i}{h}).$$

It is also reasonable to use other distributions instead of the uniform distribution with expectation zero and scale parameter h. This approach is generalized for general weight functions, which are called kernel functions.

Definition: K is kernel of order $p, p > 1$, $\int K(u)du = 1$, if and only if

$$\int u^k K(u)du = 0 \; for \; k = 1, \ldots, p-1, \quad \int u^p K(u)du = \gamma_p \neq 0.$$

5.3 Kernel Density Estimator

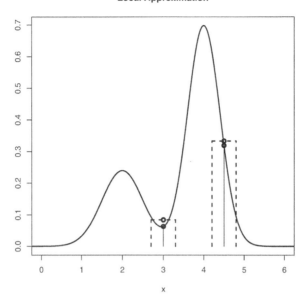

FIGURE 5.4: Local approximation.

Note that for $p = 2$ the kernel is a standardized density belonging to a family of distributions with expectation zero. A few standard kernels are plotted in Figure 5.5.

For $p > 2$, the kernel functions have negative values see the kernel functions in Figure 5.6.

Set
$$K_h(u) := \frac{1}{h} K\left(\frac{u}{h}\right).$$

The kernel density estimator is defined as
$$\widehat{f_h}(x) = \frac{1}{n} \sum_{i=1}^{n} K_h(x - X_i) = \frac{1}{nh} \sum_{i=1}^{n} K\left(\frac{x - X_i}{h}\right).$$

The estimator is also known as Parzen-Rosenblatt window or Parzen estimator. Besides the interpretation for a good estimator of a good local approximation, the kernel density estimator explores a smoothing principle. Each data point is included via a weight function, which is like lying additional noise on single points, see Figure 5.7.

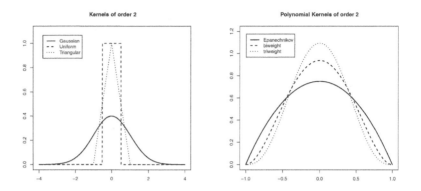

FIGURE 5.5: Left: Standard kernels. Right: Kernels of type $c_k(1-x^2)^k$ for $k = 1, 2, 3$.

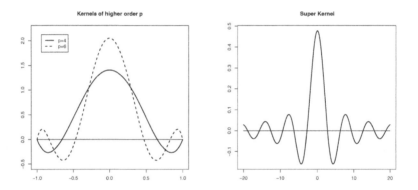

FIGURE 5.6: Optimal kernels given in Gasser et al. (1985). Right: A super kernel fulfills $\int u^k K(u) du = 0$ for all k.

5.3.1 Statistical Properties

For smooth densities, the bias and the variance of the kernel density estimator, with a symmetric kernel of order p, can be approximated by

$$Bias(\widehat{f}_h(x)) = \gamma_p \frac{h^p}{p!} f^{(p)}(x) + o(h^p), \; h \to 0,$$

where $f^{(p)}(x)$ is the p'th derivative of the density, and

$$Var(\widehat{f}_h(x)) = \frac{1}{nh} \|K\|^2 f(x) + o(\frac{1}{nh}), \; nh \to \infty,$$

where $\|K\|^2 = \int K(x)^2 dx$. For the mean squared error

$$MSE\left(\widehat{f}_h(x)\right) = E\left(\widehat{f}_h(x) - f(x)\right)^2,$$

5.3 Kernel Density Estimator

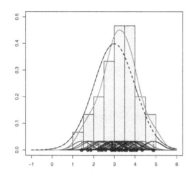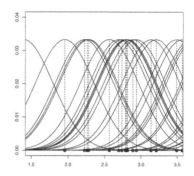

FIGURE 5.7: Left: Comparison of kernel density estimator and histogram. Right: Each observed point is blurred by a weight function.

we obtain

$$MSE\left(\widehat{f}_h(x)\right) = \frac{1}{nh}\|K\|^2 f(x) + h^{2p}\left(\frac{\gamma_p}{p!}f^{(p)}(x)\right)^2 + o(\frac{1}{nh}) + o(h^{2p}). \quad (5.2)$$

For the mean integrated squared error, using $\int f(x)dx = 1$ we get

$$MISE\left(\widehat{f}_h\right) = \frac{1}{nh}\|K\|^2 + \frac{h^{2p}}{p!}\gamma_p^2 \left\|f^{(p)}\right\|^2 + o(\frac{1}{nh}) + o(h^{2p}), \quad (5.3)$$

where $\left\|f^{(p)}\right\|^2 = \int f^{(p)}(x)^2 dx$. The leading term in (5.3)

$$AMISE\left(\widehat{f}_h\right) := \frac{1}{nh}\|K\|^2 + \frac{h^{2p}}{p!}\left\|f^{(p)}\right\|^2$$

is minimal for

$$h_{opt} \asymp n^{\frac{-1}{2p+1}}. \quad (5.4)$$

Then $MISE\left(\widehat{f}_{h_{opt}}\right) \approx n^{-\frac{2p}{2p+1}}$ attains optimal convergence rate for all densities with p continuous derivatives. Note that for $p > 2$, the kernel has negative values with the positive probability: $\widehat{f}_h(x) < 0$, for some x. Otherwise, kernels of order p deliver better convergence rates for smooth densities with derivatives of order p.

Derivation of (5.3) for smooth densities and $p = 2$:
We have

$$MSE\left(\widehat{f}_h(x)\right) = Var\left(\widehat{f}_h(x)\right) + Bias(\widehat{f}_h(x))^2$$

with

$$Bias(\widehat{f}_h(x))^2 = \left(f(x) - E\widehat{f}_h(x)\right)^2.$$

First, we consider the bias.

$$\begin{aligned}E\widehat{f}_h(x) &= \frac{1}{nh}\sum_{i=1}^n \int K\left(\frac{x-u}{h}\right) f(u)du \\ &= \int K\left(\frac{x-u}{h}\right) f(u)\frac{1}{h}du = \int K(s)\,f(x+hs)ds.\end{aligned}$$

We have changed the variable of integration $s = \frac{x-u}{h}$, $ds = \frac{1}{h}du$. Applying $f(x+hs) = f(x) + o(h)$ and $\int K(s)\,ds = 1$, we get

$$E\widehat{f}_h(x) = \int K(s)\,f(x+hs)ds = f(x) + o(h). \tag{5.5}$$

The variance term is

$$Var\left(\widehat{f}_h(x)\right) = \frac{1}{(nh)^2}\sum_{i=1}^n Var\left(K\left(\frac{x-X}{h}\right)\right) = \frac{1}{n}Var\left(\frac{1}{h}K\left(\frac{x-X}{h}\right)\right)$$

$$= \frac{1}{n}\int K\left(\frac{x-u}{h}\right)^2 f(u)\frac{1}{h^2}du - \frac{1}{n}\left(\int K\left(\frac{x-u}{h}\right) f(u)\frac{1}{h}du\right)^2 = V_1 - V_2.$$

Using $s = \frac{x-u}{h}$, $ds = \frac{1}{h}du$ and $f(x+hs) = f(x) + o(h)$ in V_1 we obtain

$$V_1 = \frac{1}{nh}\int K(s)^2\,f(x+hs)ds = \frac{1}{nh}\int K(s)^2\,ds\,f(x) + \frac{1}{nh}o(h).$$

By applying (5.5) in the second term, we obtain

$$V_2 = \frac{1}{n}(f(x) + o(h))^2 = o\left(\frac{1}{nh}\right).$$

Thus

$$Var\left(\widehat{f}_h(x)\right) = \frac{1}{nh}\int K(s)^2\,ds\,f(x) + o\left(\frac{1}{nh}\right).$$

Now, we calculate the bias term $Bias(\widehat{f}_h(x))^2$

$$\left(\int K(s)\,ds f(x) - E\widehat{f}_h(x)\right)^2 = \left(\int K(s)\,(f(x+hs) - f(x))ds\right)^2.$$

Applying the Taylor expansion of the second order

$$f(x+hs) = f(x) + hs\,f'(x) + \frac{(hs)^2}{2}f''(x) + o(h^2),$$

we get

$$Bias(\widehat{f}_h(x))^2 = \left(\int K(s)\left(hs\,f'(x) + \frac{(hs)^2}{2}f''(x) + o(h^2)\right)ds\right)^2,$$

5.3 Kernel Density Estimator

and

$$\int (K(s)(hs\, f'(x) + \frac{(hs)^2}{2} f''(x) + o(h^2))ds$$
$$= hf'(x) \int K(s)\, s\, ds + \frac{h^2}{2} f''(x) \int s^2 K(s)\, ds + o(h^2)$$
$$= \frac{h^2}{2} f''(x) \gamma_2 + o(h^2),$$

because $\int K(s)\, s\, ds = 0$, where $\gamma_2 = \int s^2 K(s)\, ds$.
Hence

$$Bias(\widehat{f}_h(x))^2 = \left(\frac{h^2}{2} f''(x) \gamma_2 + o(h^2)\right)^2.$$

After summarizing, we obtain

$$MSE\left(\widehat{f}_h(x)\right) = \frac{1}{nh} \int K(s)^2\, ds\, f(x) + \frac{h^4}{4} f''(x)^2 \gamma_2^2 + o(h^4) + o(\frac{1}{nh}). \quad (5.6)$$

For the integrated mean squared error it holds:

$$MISE\left(\widehat{f}_h(x)\right) = \int MSE\left(\widehat{f}_h(x)\right) dx$$
$$= \frac{1}{nh} \int K(s)^2\, ds + \frac{h^4}{4} \int f''(x)^2 dx\, \gamma_2^2 + o(h^4) + o(\frac{1}{nh}). \quad (5.7)$$

Derivation of (5.4) for smooth densities and $p = 2$:
Consider the leading terms of MSE in (5.6) and $MISE$ in (5.7) as functions of the bandwidth. Introduce the asymptotic mean squared error

$$AMSE(h) = \frac{1}{nh} c_1 + \frac{h^4}{4} c_2,$$

with

$$c_1 = \int K(s)^2\, ds\, f(x), \quad c_2 = f''(x)^2 \gamma_2^2.$$

We have

$$h_{AMSE} = \arg\min_h (\frac{1}{nh} c_1 + \frac{h^4}{4} c_2) = \left(\frac{c_1}{c_2} \frac{1}{n}\right)^{\frac{1}{5}}.$$

Thus, the asymptotic best bandwidth is

$$h_{AMSE} = n^{-\frac{1}{5}} \left(\frac{f(x) \int K(s)^2\, ds}{f''(x)^2 \gamma_2^2}\right)^{\frac{1}{5}}.$$

Note that the minimum is attained where the derivatives of the leading variance term and the derivative of the leading bias term are equal. Both terms of the AMSE have the same convergence rate. We get

$$MSE(h_{AMSE}) = O(n^{-\frac{4}{5}}).$$

The same calculations can be done for the $MISE$, see Figure 5.8. Define

$$AMISE(h) = \frac{1}{nh} \int K(s)^2 ds + \frac{h^4}{4} \int f''(x)^2 dx \, \gamma_2^2,$$

then

$$h_{AMISE} = \arg\min AMISE(h) = n^{-\frac{1}{5}} \left(\frac{\int K(s)^2 ds}{\int f''(x)^2 dx \, \gamma_2^2} \right)^{\frac{1}{5}}. \quad (5.8)$$

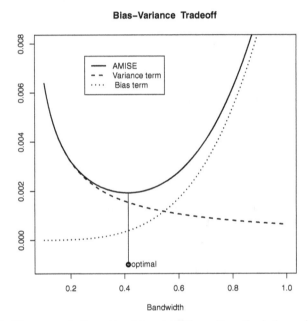

FIGURE 5.8: The asymptotic optimal bandwidth realizes the balance between variance and bias.

Now, let us discuss an optimal choice of the kernel function. The optimal bandwidth h_{AMISE} has the structure

$$h^\# = h \left(\frac{\int K(s)^2 ds}{\gamma_2^2} \right)^{\frac{1}{5}}.$$

Thus

$$AMISE(h^\#) = \left(\frac{1}{nh} + \frac{h^4}{4} \int f''(x)^2 dx \right) T(K),$$

5.3 Kernel Density Estimator

with

$$T(K) = \gamma_2^{\frac{2}{5}} \left(\int K(s)^2 \, ds \right)^{\frac{4}{5}}.$$

The kernel which minimizes $T(K)$ is the Epanechnikov kernel, defined by

$$K(s) = \frac{3}{4}(1 - s^2) I_{[-1,1]}(s),$$

and drawn in Figure 5.5.

Example 5.1 (Old Faithful) Old Faithful is a geyser in the Yellowstone National Park. On the afternoon of September 18, 1870, members of the Washburn–Langford–Doane Expedition discovered the geyser. In the report of the expedition, it is written: "It is spouted at regular intervals nine times during our stay, the columns of boiling water being thrown from ninety to one hundred feet at each discharge, which lasted from fifteen to twenty minutes. We gave it the name 'Old Faithful'". In R, two data sets related to the geyser Old Faithful are included: data(geyser), and data(faithful). Every data set contains two measurement series: eruption time (duration) in minutes and the waiting time (waiting) to the next eruption in minutes. Because the distributions of these variables are bimodal and do not follow a standard distribution, these data sets are now very popular textbook examples for smoothing techniques.

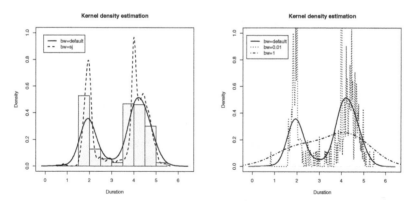

FIGURE 5.9: Left: Density estimation of eruptions duration. Right: Bandwidth=1 gives an oversmoothing of the density. The two modes are not visible . Besides, the bandwidth=0.01 gives an overfit.

R Code 5.3.28. Old Faithful, Example 5.1.

```
library(MASS); data(geyser); attach(geyser)
help(density) ### kernel density estimation
```

```
plot(density(duration),lwd=2, ylim=c(0,1),
    main=''kernel density estimation'')
lines(density(duration, bw = ''sj''),col=2)
### estimated optimal asymptotical bandwidth
lines(density(duration, bw = 0.1),col=3) ### bandwidth=0.1
density(duration,kernel=''gaussian'')
density(duration,kernel=''epanechnikov'')
```

5.3.2 Bandwidth Selection in Practice

In Figure 5.9, it is illustrated that the goodness of a kernel estimator depends essentially on the choice of the bandwidth. Below are the most common rules for a practicable bandwidth choice.

1. *Thumb rules:* The plug-in-method is based on the asymptotical optimal bandwidth h_{AMISE} in (5.8). The main principle is that independent of the applied kernel, a Gaussian kernel is plugged in (5.8); independent of the true underlying density, a normal density with expectation zero and variance σ^2 is plugged in (5.8).

    We have for $K(s) = \varphi(s)$, where φ is the density of the standard normal distribution

    $$\int K(s)^2 \, ds = \frac{1}{2\sqrt{\pi}} \int \frac{1}{\sqrt{2\pi}} \sqrt{2} \exp(-s^2) ds = \frac{1}{2\sqrt{\pi}}.$$

    Furthermore, for f equal to the density of a normal distribution with expectation zero and variance σ^2, we get

    $$f''(x) = \frac{1}{\sigma^4}(x^2 - \sigma^2) f(x),$$

    and

    $$\int f''(x)^2 dx = \frac{1}{2\sqrt{\pi}} \frac{1}{\sigma^9} E(X^2 - \sigma^2)^2,$$

    where X is normally distributed with expectation zero and variance $\frac{\sigma^2}{2}$, thus

    $$\int f''(x)^2 dx = \frac{3}{8\sqrt{\pi}} \frac{1}{\sigma^5}.$$

    If the unknown variance σ^2 is estimated by the sample variance s^2 we obtain Scott's rule,

    $$\widehat{h}_{thumb} = \left(\frac{4s^5}{3n}\right)^{\frac{1}{5}} \approx 1.06 \, s \, n^{-\frac{1}{5}},$$

5.3 Kernel Density Estimator

which is implemented in R as bw.nrd. The Silverman's rule of thumb uses a robust variance estimator. Consider the interquartile rank

$$\hat{\lambda} = x_{[0.75n]} - x_{[0.25n]},$$

where $x_{[1]}, \ldots, x_{[n]}$ is the order statistics. Under the normal distribution we have $\hat{\lambda} \approx 1.34\sigma$, such that $\tilde{\sigma} = \min(s, \frac{\hat{\lambda}}{1.34})$ can be considered as an estimator of the standard deviation. Silverman's rule of thumb for the bandwidth is

$$\hat{h}_{thumb,0} = 1.06 \min(s, \frac{x_{[0.75n]} - x_{[0.25n]}}{1.34}) n^{-\frac{1}{5}}.$$

which is implemented in R included as bw.nrd0.

2. *sj:* This method is proposed by Sheather and Jones (1991). They proposed h_{sj}, which is based on an adaptive estimation of the h_{AMISE} in (5.8). The idea is to use a pilot estimator

$$\tilde{f}_{g(h)}(x) = \frac{1}{ng(h)} \sum_{i=1}^{n} \varphi\left(\frac{x - X_i}{g(h)}\right),$$

with Gaussian kernel φ and bandwidth $g(h)$. Using partial integration twice, we get

$$\theta = \int f''(x)^2 dx = \int f^{(4)}(x) f(x) dx,$$

where $f^{(4)}(x)$ is the fourth derivative of $f(x)$. The fourth derivative $f^{(4)}(x)$ is estimated by the fourth derivative of $\tilde{f}_{g(h)}(x)$:

$$\tilde{f}^{(4)}_{g(h)}(x) = \frac{1}{ng(h)^5} \sum_{i=1}^{n} \varphi^{(4)}\left(\frac{x - X_i}{g(h)}\right).$$

The integral is estimated by the average. Thus

$$\hat{\theta} = \hat{S}(g(h)) =$$

$$\frac{1}{n-1} \sum_{j=1}^{n} \tilde{f}^{(4)}_{g(h)}(x_j) = \frac{1}{n(n-1)} \frac{1}{g(h)^5} \sum_{j=1}^{n} \sum_{i=1}^{n} \varphi^{(4)}\left(\frac{X_j - X_i}{g(h)}\right).$$

The bandwidth $g(h)$ of the pilot estimate is an adaptive optimal bandwidth for estimating θ, and is given by

$$g(h) = 1.357 h^{\frac{5}{7}} \left(\frac{\hat{S}(a)}{\hat{T}(b)}\right)^{\frac{1}{7}},$$

where $\widehat{T}(b)$ is the estimate of $\int f^{(3)}(x)^2 dx$

$$\widehat{T}(b) = \frac{-1}{n(n-1)} \frac{1}{b^7} \sum_{j=1}^{n} \sum_{i=1}^{n} \varphi^{(6)}\left(\frac{X_j - X_i}{b}\right).$$

It is proposed that $a = 0.920\widehat{\lambda} n^{-\frac{1}{7}}$ and $b = 0.912\widehat{\lambda} n^{-\frac{1}{9}}$ where $\widehat{\lambda}$ is the sample interquartile range. Finally, the bandwidth of h_{sj} of the kernel density estimator is the solution of

$$h = n^{-\frac{1}{5}} \left(\frac{\int K(s)^2 ds}{\widehat{S}(g(h))\, \gamma_2^2}\right)^{\frac{1}{5}},$$

calculated with the Newton-Raphson method. Note that the procedure includes three different levels of bandwidth estimators: $h, g(h), a$ and b.

3. *Maximum likelihood cross-validation:* Consider the Kullback Leibler (KL) information as the distance measure between f and its estimator $\widehat{f}_h$:

$$d_{KL}\left(f, \widehat{f}_h\right) = E_f \ln\left(\frac{f}{\widehat{f}_h}\right).$$

The main idea is to choose the optimal bandwidth such that $d_{KL}\left(f, \widehat{f}_h\right)$ is minimized, meaning that $E_f \ln(\widehat{f}_h)$ is maximized. The cross-validation estimator $CV_{KL}(h)$ of $E_f \ln(\widehat{f}_h)$ is

$$CV_{KL}(h) = \frac{1}{n} \sum_{i=1}^{n} \ln \widehat{f}_{h,i}(X_i),$$

where $\widehat{f}_{h,i}(x)$ is the leave-one-out estimate

$$\widehat{f}_{h,i}(x) = \frac{1}{(n-1)} \sum_{j \neq i} K_h(x - X_j). \tag{5.9}$$

Then, the bandwidth is chosen such that

$$\widehat{h}_{KL} = \arg\max_{h} CV_{KL}(h).$$

4. *Least squares cross-validation:* Consider the integrated squares error (ISE) as the distance measure between f and its estimator $\widehat{f}_h$

$$ISE\left(\widehat{f}_h\right) = \int (f - \widehat{f}_h)^2 dx.$$

5.5 Orthogonal Series Estimator

The main idea is to choose the optimal bandwidth such that $ISE\left(\widehat{f}_h\right)$ is minimized or respective

$$\int \widehat{f}_h(x)^2 dx - 2\int f(x)\widehat{f}_h(x)dx$$

is minimized. $\int f(x)\widehat{f}_h(x)dx$ is estimated by the cross-validation method: $\frac{1}{n}\sum_{i=1}^n \widehat{f}_{h,i}(X_i)$, where $\widehat{f}_{h,i}(x)$, is the leave-one-out estimate given in (5.9). Then, the choice of bandwidth is

$$\widehat{h}_{CV} = \arg\min_h CV_{LSE}(h),$$

with

$$CV_{LSE}(h) = \int \widehat{f}_h(x)^2 dx - \frac{2}{n}\sum_{i=1}^n \widehat{f}_{h,i}(X_i).$$

5.4 Nearest Neighbor Estimator

These estimators are similar to the kernel estimator; however, the bandwidth is determined by the number of observations inside the neighborhood. Let k_n be given such that $k_n/n \to 0$ and $k_n \to \infty$. Define $R(x)$ the distance of x to the k_n closest of $X_1, \ldots, X_n$, then

$$\widehat{f}_h(x) = \frac{1}{nR(x)}\sum_{i=1}^n K\left(\frac{x - X_i}{R(x)}\right).$$

5.5 Orthogonal Series Estimator

Suppose $f \in \mathcal{L}(e_1, \ldots, e_k, \ldots)$ such that $f(x) = \sum_k a_k e_k(x)$, $\{e_k\}$ an orthogonal basis with $\int e_k(x) e_j(x) w(x) dx = \delta_{kj}$. Orthogonal series estimation has two main aspects: cutting the series for all $k \notin K(x)$, and estimation of the coefficients f_k. An orthogonal series estimator is given by

$$\widehat{f}_K(x) = \sum_{k \in K(x)} \widehat{a}_k\, e_k(x), \quad \widehat{a}_k = \frac{1}{n}\sum_{i=1}^n e_k(X_i)\, w(X_i).$$

5.6 Minimax Convergence Rate

In this section, we will give some insights on the convergence rate of MISE. For the histogram, we obtained $MISE\left(\widehat{f}_{h_0}\right) \approx n^{-\frac{2}{3}}$, where h_0 is the asymptotically optimal bandwidth.

For the kernel density estimator $\widehat{f}_{h_{opt}}$ we have $MISE\left(\widehat{f}_{h_{opt}}\right) \approx n^{-\frac{2p}{2p+1}}$ in case that $f \in \mathcal{F}_p$, where

$$\mathcal{F}_p = \{f : \mathbb{R} \to \mathbb{R}, f(x) \geq 0, \int f(x)dx = 1,$$

there exist continuous derivatives, $f^{(k)}, k = 1, \ldots, p\}$,

and that the kernel is of order p.

In case of a parametric setup, meaning that we have an i.i.d. sample $X_1, \ldots, X_n$ from P_θ, $\theta \in \mathbb{R}^k$, we know that under regularity conditions, the best possible convergence rate of MSE is n^{-1}. Why do we have slower rates for kernel estimators and histograms? For explaining the background, we follow the argumentation of Hall (1989).

Consider the model

$$\mathcal{L} = \{f : [0,1] \to \mathbb{R}, f(x) \geq \epsilon > 0, \int f(x)^2 dx < \infty,$$

$$f(x) = \sum_{j=0}^{\infty} a_j \cos(j\pi x), |a_j| \leq c_j\}$$

for some constants c_j and some ϵ.

Let $\widehat{f} \in \mathcal{L}$ be an arbitrary estimator of $f \in \mathcal{L}$, then

$$\widehat{f}(x) = \sum_{j=1}^{\infty} \widehat{a}_j \cos(j\pi x),$$

with

$$\widehat{a}_0 = \int_0^1 \widehat{f}(x)dx, \quad \widehat{a}_j = 2\int_0^1 \widehat{f}(x)\cos(j\pi x)dx.$$

Using Parseval's equality, we obtain

$$E\left(\int_0^1 (f(x) - \widehat{f}(x))^2 dx\right) = E\left((\widehat{a}_0 - a_0)^2 + \frac{1}{2}\sum_{j=1}^{\infty} (\widehat{a}_j - a_j)^2\right).$$

Hence, we have a lower bound

$$MISE \geq \frac{1}{2}\sum_{j=1}^{\infty} E(\widehat{a}_j - a_j)^2 = \frac{1}{2}\sum_{j=1}^{\infty} \left(Var(\widehat{a}_j) + (\alpha_j - a_j)^2\right),$$

5.6 Minimax Convergence Rate

where $\alpha_j = E\widehat{a}_j$. Let α_j be a smooth function of a_j with derivative α'_j. The Cramer-Rao bound applied to every Fourier coefficient estimate $\widehat{a}_j$, and it gives

$$Var(\widehat{a}_j) \geq \frac{(\alpha'_j)^2}{I_{(n)}},$$

where $I_{(n)}$ is the Fisher information

$$I_{(n)} = \sum_{i=1}^{n} E\left(\frac{d\ln f(X_i)}{da_j}\right)^2.$$

For $f(x) = \sum_{j=0}^{\infty} a_j \cos(j\pi x)$, it holds

$$\frac{d\ln f(x)}{da_j} = \frac{1}{f(x)}\frac{df(X_i)}{da_j} = \frac{\cos(j\pi x)}{f(x)}.$$

Because $f(x) \geq \epsilon$ on $[0,1]$, we get

$$I_{(n)} = nE\left(\frac{\cos(j\pi x)}{f(x)}\right)^2 < \frac{n}{\epsilon}\int \cos(j\pi x)^2 dx = \frac{n}{2\epsilon}.$$

Thus

$$MISE \geq \frac{1}{2}\sum_{j=1}^{\infty}\left(\frac{(\alpha'_j)^2 2\epsilon}{n} + (\alpha_j - a_j)^2\right).$$

Considering a minimax bound, the best possible estimator for the most complicated estimation problem inside the given model, we have

$$\inf_{\widehat{f}} \sup_{f \in \mathcal{L}} MISE \geq \frac{\epsilon}{2}\sum_{j=1}^{\infty} \inf_{\alpha_j} \sup_{|a_j| < c_j}\left(\frac{2(\alpha'_j)^2}{n} + (\alpha_j - a_j)^2\right).$$

There exists a useful auxiliary result for all continuously differentiable functions $g: [a,b] \to \mathbb{R}$ and some positive constant δ:

$$\sup_{x \in [a,b]}\left(\delta g'(x)^2 + (g(x) - x)^2\right) \geq \frac{(b-a)^2 \delta}{(b-a)^2 + 4\delta}.$$

The proof can be found in Hall (1989). We apply the above inequality on each term of the sum, where α_j is a function of a_j, the interval length is c_j, and $\delta = \frac{2}{n}$. We obtain

$$\inf_{\widehat{f}} \sup_{f \in \mathcal{L}} MISE \geq \frac{\epsilon}{2}\sum_{j=1}^{\infty}\frac{2c_j^2 n^{-1}}{c_j^2 + 8n^{-1}} = \frac{\epsilon}{2} rate,$$

with

$$rate = \sum_{j=1}^{\infty}\frac{c_j^2}{c_j^2 n + 8}.$$

We take a closer look at the rate. For functions $f \in \mathcal{F}_p$, where the pth derivative $f^{(p)}$ fulfills the Hölder condition with $\alpha = \frac{1}{2}$, it holds $c_j = cj^{-(p+\alpha)}$. For more details, see the textbook Zygmund (2003), page 71. Take the integer number N with $N \leq n^m \leq N+1$ where $m = \frac{1}{2p+1}$ and split the sum, thus

$$rate = \sum_{j=1}^{N} \frac{c_j^2}{c_j^2 n + 8} + \sum_{j=N+1}^{\infty} \frac{c_j^2}{c_j^2 n + 8} = rate_{(1)} + rate_{(2)}.$$

We can consider the first term responsible for the quality of estimation of the first N coefficients. The second term is for the approximation. For smooth functions, the Fourier coefficients are quickly decreasing. The coefficients of order higher than $N+1$ are included in the approximation error. Note that $N \geq \frac{N+1}{2} \geq \frac{1}{2}n^m$. Then

$$rate_{(1)} = \sum_{j=1}^{N} \frac{c_j^2}{2c_j^2 n + 8} \geq N \frac{1}{n} \min_{j: j \leq N} \frac{c_j^2}{c_j^2 + 8n^{-1}} \geq \text{const } n^m n^{-1} = \text{const } n^{-\frac{2p}{2p+1}}.$$

Because $N \leq n^m$ and $\min_{j \leq N} c_j^2 \geq \text{const } N^{-(2p+1)} \geq \text{const } n^{-1}$, we get

$$\min_{j \leq N} \frac{c_j^2}{c_j^2 + 8n^{-1}} = \frac{\min_{j \leq N} c_j^2}{\min_{j \leq N} c_j^2 + 8n^{-1}} \geq const.$$

Now we consider the second term

$$rate_{(2)} = \sum_{j=N+1}^{\infty} \frac{c_j^2}{c_j^2 n + 8}.$$

We have $n^m \leq N+1$ and

$$\max_{j, j \geq N+1} (c_j^2 n + 8) \leq \text{const } (N+1)^{-(2p+1)} n + 8 \leq const.$$

Thus

$$rate_{(2)} \geq \text{const} \sum_{j=N+1}^{\infty} c_j^2 \geq N^{-(2p+1)} \sum_{j=1}^{\infty} cj^{-(2p+1)} \geq \text{const } N^{-(2p+1)} \geq \text{const } n^{-1}.$$

After summarizing, we get

$$rate = \text{const}(n^{-\frac{2p}{2p+1}} + n^{-1}) \geq \text{const} n^{-\frac{2p}{2p+1}}.$$

These calculations illustrate the balance of approximation and estimation. Estimating a function requires the estimation of infinitely many parameters. However, estimating a smooth function only requires estimation of the slowly increasing number of parameters.

5.7 Problems

1. Consider a kernel density estimator $\widehat{f}(x)$ with Epanechnikov kernel $(K(s) = \frac{3}{4}(1 - s^2)I_{[-1,1]}(s);\ \int K(s)^2 ds = \int_{-1}^{1} \left(\frac{3}{4}\right)^2 (1 - s^2)^2 ds = \frac{3}{5},\ \int s^2 K(s)ds = \int_{-1}^{1} s^2 \left(\frac{3}{4}\right)(1 - s^2)ds = \frac{1}{5}).$

   (a) Derive the leading term for the bias term $(E\widehat{f}(x) - f(x))^2$.
   (b) Derive the leading term for the variance $Var(\widehat{f}(x))$.
   (c) Discuss the consistency of the estimator.
   (d) Suppose that the density of $N(1,1)$ is a good start estimator for the unknown density $f(x)$. Give a recommendation for an asymptotic optimal bandwidth at $x = 1$ and $x = 2.64$.
   (e) Explain why it is reasonable to choose a local bandwidth larger than the other.

2. Simulate several samples of a normal distribution and perform Shapiro test of normality.

   (a) Estimate the density of the test statistics.
   (b) Estimate the density of the p-values.
   (c) Are the p-values uniformly distributed? (use chi-squared test, ks-test, and qq-plot).

3. Simulate several samples of a "non-normal" distribution and perform Shapiro test of normality.

   (a) Estimate the density of the test statistics.
   (b) Estimate the density of the p-values.
   (c) Are the p-values uniformly distributed? (use chi-squared test, ks-test, and qq-plot).

4. Given the following R code for a test problem

   ```
   ### data set simulation
   p <- rbinom(100,1,0.3)
   y <- rnorm(100,-2,1)*p+rnorm(1000,2,1)*(1-p)
   fhat <- density(y,n=100,from=-6,to=6); xx<-seq(-6,5.88,0.12)
   ftrue <- dnorm(xx,-2,1)*0.2+dnorm(xx,2,1)*(1-0.2)

   ### Ho: 0.5*N(-2,1)+0.5*N(2,1)
   xx <- seq(-6,5.88,0.12)
   f0 <- dnorm(xx,-2,1)*0.5+dnorm(xx,2,1)*0.5
   tobs <- sum(abs(f0-fhat$y))/length(xx)

   #### plot
   plot(xx,f0,ylim=c(0,0.35),''l'',lwd=2, main=''Examination'')
   ```

```
lines(xx,ftrue,''l'',lwd=2); lines(fhat,col=2,lty=2,lwd=2)
p0 <- rbinom(1000,1,0.4)
y0<-rnorm(1000,-2,1)*p0+rnorm(100,2,1)*(1-p0)
f <- density(y0,n=100,from=-6,to=6); lines(f,col=3,lty=3,lwd=2)

### test
B <- 100
T <- rep(NA,B)
for(j in 1:B){
    p0<-rbinom(100,1,0.5);
    y0<-rnorm(100,-2,1)*p0+rnorm(100,2,1)*(1-p0)
    f<-density(y0,n=100,from=-6,to=6)
    T[j]<-sum(abs(f$y-f0))/100}
pvalue <- length(T[T>tobs])/B
```

$t_{obs} = 0.03$, $pvalue = 0$

(a) Formulate the test problem.

(b) Which test statistic is used?

(c) How is the p-value defined? How is the p-value calculated?

(d) Does this algorithm work? What is the test result?

6
Nonparametric Regression

In this chapter, we discuss regression analysis, a powerful method for examining the relationship between two or more variables.

6.1 Background

Let us begin with a general introduction to regression. Consider a two variate i.i.d. sample $((X_1, Y_1), \ldots, (X_n, Y_n))$ from (X, Y) with density $f_{(X,Y)}(x, y)$ and respective marginal densities $f_X(x)$, $f_Y(y)$. The Y variable is called the response or the dependent variable, and the X variable is called the covariate or the independent variable. The goal is to explain the response Y with the help of a function of X, such that the mean square error is minimal:

$$\min_{g \in \mathcal{G}} E\left(Y - g(X)\right)^2.$$

Note that the expectation is taken with respect to both variables (X, Y). Define the conditional mean

$$m(x) = E(Y \mid X = x) = \int y \frac{f_{(X,Y)}(x, y)}{f_X(x)} dy, \qquad (6.1)$$

then

$$E\left(Y - g(X)\right)^2 = E\left(Y \pm m(X) - g(X)\right)^2$$
$$= E\left(Y - m(X)\right)^2 + E\left(m(X) - g(X)\right)^2,$$

because of the projection property of the conditional expectation

$$E\left((Y - m(X))(m(X) - g(X))\right) = E\left((m(X) - g(X))E\left((Y - m(X)) \mid X\right)\right) = 0.$$

This means $\min_{g \in \mathcal{G}} E\left(Y - g(X)\right)^2 = E\left(Y - m(X)\right)^2$. The goal is therefore to estimate the conditional expectation $m(x)$.

We interpret the conditional mean as an expectation of Y when the respective partner X lies in a neighborhood of x. We can see two contractive principles of estimation $m(x)$:

- Estimate the expectation by averaging. The more variables included, the better the estimate.

- Determine the local environment of x. If the area is smaller, the approximation will be better, but fewer variables will be included in the average.

The nonparametric regression techniques are weighted averages of the response Y:

$$\widehat{m}_h(x) = \sum_{i=1}^{n} W_{hi}(x) Y_i, \quad W_{hi}(x) = W_{hi}(x, X_1, \ldots, X_n).$$

This approach is mainly explored by the kernel regression estimators. The nonparametric regression model is often formulated by the equation

$$Y_i = m(x_i) + \varepsilon_i, \ i = 1, \ldots, n, \tag{6.2}$$

where the nonobserved $(\varepsilon_1, \ldots, \varepsilon_n)$ are i.i.d. with expected value zero and finite variance. The distribution of the error term ε_i is independent of x_i. For illustration, see Figure 6.1.

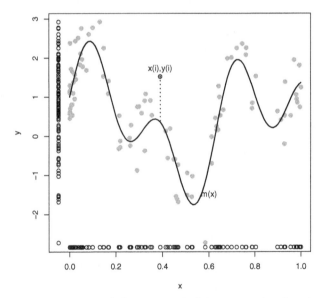

FIGURE 6.1: Data points, and the true underlying regression function $m(x)$. Illustration of Equation 6.2.

The formulation (6.2) is also useful for fixed design regressions where Y_i is observed at given fixed design points x_i, $i = 1, \ldots, n$. In this case, the data is not an i.i.d. sample. The $Y_1, \ldots, Y_n$ are independent, not identically distributed.

6.1 Background

Similar to Chapter 5, we assume no parametric formula for m. We require that

$$m \in \mathcal{S}^k = \{m : \mathbb{R} \to \mathbb{R}, m \text{ is continuously differentiable up to order } k,$$
$$\text{and } \int (m^{(k)})^2 dx < \infty\},$$

then we search for a function $\widehat{m}$ in $\mathcal{G}$, where

$$\mathcal{G} = \{\widehat{m} : \mathbb{R} \to \mathbb{R}, \widehat{m}(x) = \widehat{m}(x, x_1, Y_1, \ldots, x_n, Y_n),$$
$$\text{measurable in } Y_1, \ldots, Y_n,$$
$$\text{and } \widehat{m}(., x_1, Y_1, \ldots, x_n, Y_n) \in \mathcal{S}^k\}.$$

This problem is similar to estimating density, the data consists of n observed pairs $(x_1, y_1), \ldots, (x_n, y_n)$, and we want to estimate $m(x)$ for all $x \in \mathbb{R}$. Therefore, we also apply here the balance between

- Approximation.

- Estimation of the approximated function.

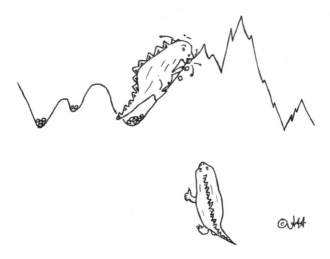

FIGURE 6.2: Principle of smoothing. Mollifiers at work.

Avoiding the approximation step supports estimators which go through all observation points $(x_1, y_1), \ldots, (x_n, y_n)$. This is the most extreme case of undersmoothing. Otherwise, a very rough approximation disregards the pattern of the underlying regression function $m(x)$, which is the oversmoothing effect. Compare the left and the right plots in Figure 6.3.

The approximated regression function can often expressed as a linear combination of known basis functions. Therefore, methods of the classical linear regression can be applied. In Section 6.4, we present restricted linear regression, followed by splines and wavelet estimators.

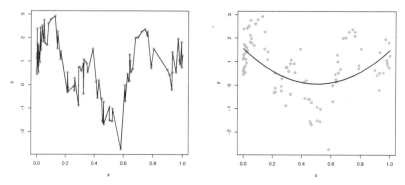

FIGURE 6.3: Left: Undersmoothing. Right: Quadratic approximation delivers an oversmoothing.

6.2 Kernel Regression Smoothing

Kernel methods in regression are strongly related to kernel density estimation. We consider the case of random X-variables and assume that the data points are observation from an i.i.d. sample $((X_1, Y_1), \ldots, (X_n, Y_n))$ from (X, Y). Then, the regression function in (6.2) is the conditional expectation

$$m(x) = E(Y \mid X = x) = \int y \frac{f_{(X,Y)}(x,y)}{f_X(x)} dy.$$

We estimate the joint density by a two-dimensional kernel density estimator

$$\widehat{f}_{(X,Y)}(x,y) = \frac{1}{nh_x h_y} \sum_{i=1}^{n} K_{x,h_x}(x - x_i) K_{y,h_y}(y - y_i),$$

and the marginal density of X is estimated in a similar way by

$$\widehat{f}_X(x) = \frac{1}{nh} \sum_{i}^{n} K_{x,h_x}(x - x_i),$$

where K_x and K_y are kernel functions. For illustration, we revisit Example 5.1.

Example 6.1 (Old Faithful) The data set faithful in R includes two variables named "waiting" and "eruptions". The variable "waiting" contains the time in minutes between two eruptions, "eruptions" is the time of the prior eruption. We are interested in forecasting the time of the next eruption. Thus, the response variable is "waiting" and the independent X variable is "eruptions".

For completeness, we give the R code for a two-dimensional density estimation, see Figure 6.4.

6.2 Kernel Regression Smoothing

R Code 6.2.29. Old Faithful, density estimation, Example 6.1.

```
library(MASS)
data(faithful)
attach(faithful)
kde2 <- kde2d(eruptions,waiting,n=64)
persp(kde2$x,kde2$y,kde2$z, col=gray(0.8), theta=30, phi=30,
      xlab="eruptions", ylab="waiting", zlab="density)
contour(kde2$x,kde2$y,kde2$z, xlab="eruptions", ylab="waiting")
```

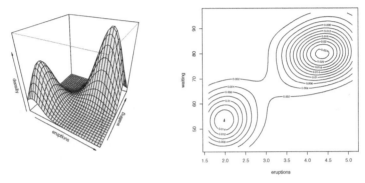

FIGURE 6.4: Left: Two-dimensional density estimation in a perspective graph. Right: The same estimate illustrated in a contour plot.

The conditional expectation $m(x)$ is estimated by plugging in the respective density estimates, see Figure 6.5. We have

$$\int y \frac{\widehat{f}_{X,Y}(x,y)}{\widehat{f}_X(x)} dy = \frac{1}{\widehat{f}_X(x)} \int y \frac{1}{nh_x h_y} \sum_{i=1}^n K_{x,h_x}(x-x_i) K_{y,h_y}(y-y_i) dy$$

$$= \frac{1}{\widehat{f}_X(x)} \frac{1}{nh_x} \sum_i K_{x,h_x}(x-x_i) \int y K_{y,h_y}(y-y_i) \frac{1}{h_y} dy.$$

Using the properties of a kernel, we get

$$\int y K_{y,h_y}(y-y_i) \frac{1}{h_y} dy = \int (y_i + h_y s) K_y(s) ds$$

$$= \int y_i K_y(s) ds + h_y \int s K(s) ds = y_i.$$

Thus

$$\int y \frac{\widehat{f}_{X,Y}(x,y)}{\widehat{f}_X(x)} dy = \frac{1}{\widehat{f}_X(x)} \frac{1}{nh_x} \sum_i^n K_{x,h_x}(x-x_i) y_i = \widehat{m}(x).$$

Note that only the kernel $K_{x,h}$ and the bandwidth h_x of the density estimation of the X-variable remains in the formulary of $\widehat{m}(x)$. Let us denote them now by K and h. The kernel regression estimation $\widehat{m}(x)$ is called Nadaraya-Watson estimation and is given by

$$\widehat{m}_h(x) = \frac{\sum_i^n K_h(x - x_i)y_i}{\sum_i^n K_h(x - x_i)}.$$

The estimation is a weighted sum of the response variable

$$\widehat{m}_h(x) = \sum_{i=1}^n W_{hi}(x)y_i,$$

where the weights are

$$W_{hi}(x) = W_{hi}(x, x_1, \ldots, x_n) = \frac{K_h(x - x_i)}{\sum_{j=1}^n K_h(x - x_j)}.$$

Note that $\sum_{i=1}^n W_{hi}(x) = 1$, for all x.

The bandwidth h is the tuning parameter. For large bandwidth, the local neighborhood around x is large and many of the y_i are averaged. This means a rough approximation with a high bias, and a good estimation with a small variance. Large bandwidth supports oversmoothing, and small bandwidth supports undersmoothing. The smoothing principle is illustrated in Figure 6.2. The Nadaraya-Watson estimator is implemented in several packages in R.

R Code 6.2.30. Old Faithful, kernel regression, Example 6.1.

```
library(MASS)
data(faithful)
attach(faithful)
plot(eruptions,waiting,lwd=2)
lines(ksmooth(eruptions,waiting,kernel="normal", bandwidth=5),lwd=2)
lines(ksmooth(eruptions,waiting,kernel="normal"),lwd=2,lty=2)
```

Plots generated by the R Code 6.2.30 are given in Figure 6.7. For further discussion of different kernel functions and the choice of bandwidth, revisit Chapter 5. Using the methods in Chapter 5, the mean squared error

$$MSE(\widehat{m}_h(x)) = E(m(x) - \widehat{m}_h(x))^2 = Var(\widehat{m}_h(x)) + Bias(\widehat{m}_h(x))^2$$

can be approximated by

$$Var(\widehat{m}_h(x)) = \frac{1}{nh}\frac{\sigma^2(x)}{f(x)}\|K\|^2 + o(\frac{1}{nh}),$$

where $\sigma^2(x) = Var(Y \mid X)$ and

$$Bias(\widehat{m}_h(x))^2 = \frac{h^4}{4}\gamma_2^2\left(m''(x) + 2\frac{m'(x)}{f(x)}f'(x)\right)^2 + o(h^4),$$

where m' and m'' are the derivatives.

The expansion shows the trade-off between variance and bias term. A large bandwidth implies a large bias term and a small variance implies oversmoothing. A small bandwidth gives a smaller bias and a higher variance.

6.3 Local Regression

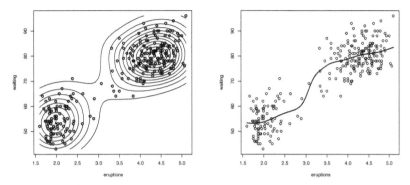

FIGURE 6.5: Left: The data points together with two contour plots of the two-dimensional kernel density estimate. Right: The same density data points and the estimated conditional expectation.

6.3 Local Regression

The Nadaraya-Watson estimator at x is a weighted average of the response inside a local neighborhood of x. This means that the underlying local approximation is a constant. In this section, we consider methods where the local approximation is a polynomial. Local regression consists of the following steps:

1. A local neighborhood is defined around x.
2. The regression function is approximated by a polynomial.
3. The coefficients of the polynomial are estimated.
4. The estimator $\widehat{m}(x)$ is the estimated polynomial at x.

Now we discuss local linear approximation and weighted least squares as estimating method. Suppose we have observed points $(x_1, y_1), \ldots, (x_n, y_n)$

$$y_i = m(x_i) + \varepsilon_i,$$

where $\varepsilon_1, \ldots, \varepsilon_n$ unobserved realizations from an i.i.d. sample with the expectation zero and finite variance.

Local regression works as well in the case of fixed design points $x_1, \ldots, x_n$. Assume we have a weight function $w_i(x) = w(x, x_i)$ defining the local neighborhood around x.

The weights can be defined with help of kernel functions $w_i(x) = K\left(\frac{x_i - x}{h}\right)$, see Figures 6.6 and 6.10. Alternatively, the weights can be determined by

$$w_{ki}(x) = \begin{cases} \frac{n}{k} & \text{if} \quad i \in J_x \\ 0 & \text{otherwise} \end{cases},$$

$J_x = \{i : X_i \text{ is one of the } k \text{ nearest observations to } x\}.$

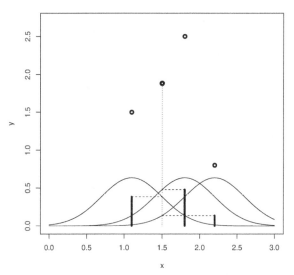

FIGURE 6.6: Principle of weighted averaging with 3 data points: We are interested in the estimate at 1.5. The weights for y_1, y_2, y_3 are defined by the height of the weight functions around x_1, x_2, x_3 variables at 1.5.

The local linear estimator is given by

$$\widehat{m}_{local}(x) = \widehat{\beta}_{0,local},$$

where

$$\left(\widehat{\beta}_{0,local}, \widehat{\beta}_{1,local}\right)^T = \arg\min_{\beta_0, \beta_1} \sum_{i=1}^{n} w_i(x) \left(y_i - \beta_0 - \beta_1(x_i - x)\right)^2.$$

See Figure 6.8. Hence

$$\widehat{m}_{local}(x) = \overline{y}_w - \frac{m_{xy}}{m_{xx}}(\overline{x}_w - x),$$

with

$$\overline{x}_w = \frac{1}{n}\sum_{i=1}^{n} w_i x_i, \quad \overline{y}_w = \frac{1}{n}\sum_{i=1}^{n} w_i y_i, \quad w_i = w_i(x),$$

and

$$\frac{m_{xy}}{m_{xx}} = \frac{\sum_{i=1}^{n} w_i (y_i - \overline{y}_w)(x_i - \overline{x}_w)}{\sum_{i=1}^{n} w_i (x_i - \overline{x}_w)^2}.$$

The estimator $\widehat{m}_{local}(x)$ is a linear function of $(y_1, \ldots, y_n)$. We can rewrite

$$\widehat{m}_{local}(x) = \sum_{i=1}^{n} l_i(x) y_i,$$

6.3 Local Regression

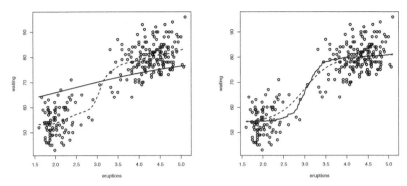

FIGURE 6.7: Left: The same Gaussian kernel but different bandwidths. The connected line is the estimator with a bandwidth of 5, which implies an oversmoothing effect. The broken line is based on the default bandwidth. Right: The same bandwidth 2 but with different kernel functions. The broken line is based on the Gaussian kernel, and the connected line is based on the box kernel function.

with
$$l_i(x) = w_i - \frac{w_i}{m_{xx}}(x_i - \overline{x}_w)(\overline{x}_w - x).$$

It holds
$$\sum_{i=1}^n l_i(x) = 1, \quad \sum_{i=1}^n l_i(x)x_i = x. \quad (6.3)$$

The main argument for local linear estimators is the benefit of bias reduction. We have
$$E\widehat{m}_{local}(x) = E\sum_{i=1}^n l_i(x)y_i = \sum_{i=1}^n l_i(x)m(x_i).$$

Plugging in the Taylor expansion for m
$$m(\xi_i) = m(x) + m'(x)(x_i - x) + rest_i(x),$$

with
$$rest_i(x) = \frac{1}{2}m''(x + \Delta_i)(x_i - x)^2,$$

we get from (6.3)
$$E\widehat{m}_{local}(x)$$
$$= \sum_{i=1}^n l_i(x)m(x) + m'(x)\sum_{i=1}^n l_i(x)(x_i - x) + \sum_{i=1}^n l_i(x)rest_i(x) \quad (6.4)$$
$$= m(x) + \sum_{i=1}^n l_i(x)rest_i(x).$$

In the case of local polynomial estimators, the bias reduction is still stronger, because the next terms in the Taylor expansion vanish.

172   6 Nonparametric Regression

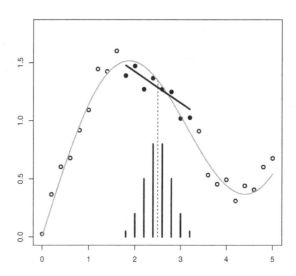

FIGURE 6.8: Principle of local linear regression.

R Code 6.3.31. Old Faithful, local regression, Example 6.1.

```
library(MASS)
library(KernSmooth)
data(faithful)
attach(faithful)
plot(eruptions,waiting)
lines(locpoly(eruptions,waiting,degree=1,bandwidth=0.5))
lines(locpoly(eruptions,waiting,degree=3,bandwidth=0.5))
```

Plots of R Code 6.3.31 are given in Figure 6.9. The local regression methods are very useful for estimating the derivatives of the regression function. Approximation by a polynomial of degree k provides the possibility to estimate the derivatives up to order $k - 1$. This is explored in the following example.

Example 6.2 (Methuselah) In the paper by Schöne et al. (2005), the data set of the bivalve (*Arctica islandica*, Mollusca; Iceland) named Methuselah is published. This 374–year old bivalve is one of the oldest reported animals. It was collected alive during July 1868 near Iceland, the exact location is unknown, see illustration in Figure 6.11. Now, it belongs to the Möbius collection at Zoological Museum Kiel, Germany. The shell of *Arctica islandica* has distinct annual and daily growth lines. The Methuselah data set includes the year (time) and the annual increments width (m).

The R code for generating Figure 6.12 and Figure 6.13 are given here.

6.4 Classes of Restricted Estimators

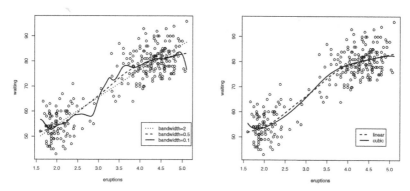

FIGURE 6.9: Old Faithful. Left: Local linear regression with the same Gaussian kernel but with a different bandwidth. Right: Gaussian kernel with bandwidth 0.5. Local polynomial regression works well for sparse regions, but there are boundary problems.

R Code 6.3.32. Methuselah data, local regression, Example 6.2.

```
library(KernSmooth)
plot(time,m,xlab="year",ylab="annual increment width (mm)" )
lines(locpoly(time,m,degree=2,bandwidth=10))
locpoly(time,m,degree=3,bandwidth=15,drv=1)
```

6.4 Classes of Restricted Estimators

As mentioned before, the estimation methods for the regression function $m(x)$ have two main aspects: approximation and estimation.

Suppose we approximate the regression function $m(x)$ by a function that lies in the subspace $\mathcal{L} = \mathcal{L}(g_1, \ldots, g_M)$ of $\mathcal{S}^k$,

$$\mathcal{L} = \{g : \mathbb{R} \to \mathbb{R}, \ g(x, \beta) = \sum_{k=1}^{M} \beta_k g_k(x), \beta = (\beta_1, \ldots, \beta_M) \in \mathbb{R}^M\}.$$

This means that we search for a good estimator of the best possible approximation defined by Figure 6.14

$$\min_{\beta \in \mathbb{R}^M} \int (m(x) - g(x, \beta))^2 dx,$$

Because $g(x, \beta)$ is a linear function in β, we can apply methods of linear regression. Suppose a surrogate linear regression model

$$\mathbf{Y} = \mathbf{X}\beta + \epsilon, \tag{6.5}$$

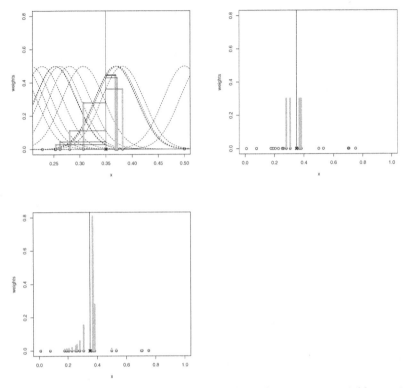

FIGURE 6.10: Left: Weights defined by kernel. Right: 5-nearest-neighbor weights. Bottom: Weights depending on the distance.

with

$$\mathbf{X} = \begin{pmatrix} g_1(x_1) & \cdots & g_k(x_1) & \cdots & g_M(x_1) \\ \vdots & & \vdots & & \vdots \\ g_1(x_i) & \cdots & g_k(x_i) & \cdots & g_M(x_i) \\ \vdots & & \vdots & & \vdots \\ g_1(x_n) & \cdots & g_k(x_n) & \cdots & g_M(x_n) \end{pmatrix}.$$

From the theory of linear models, we know that the least squares estimator is the best linear unbiased estimator in (6.5). Because we are estimating an approximation of the regression function m, the unbiasedness is no longer our greatest interest. Biased estimators can improve the mean squared error of the least squares estimator. One important class is the restricted regression estimators. The main idea is to minimize the least squares distance between the observations and a fitted value inside

$$\mathcal{L}(\mathbf{X}) = \left\{ \mathbf{X}\beta : \beta \in \mathbb{R}^M \right\}$$

6.4 Classes of Restricted Estimators

FIGURE 6.11: A 374-year-old Methuselah mollusk was collected alive in July 1868 near Iceland. The finder and the location are unknown.

and penalize the choice of large coefficients β_j by a penalizing term $pen(\beta)$:

$$C(\beta, \lambda) = (Y - X\beta)^T(Y - X\beta) + \lambda pen(\beta).$$

The scale parameter λ is a tuning parameter that controls the strength of the penalty. Keep in mind that the original problem is to estimate the regression function m; the approximation of m inside $\mathcal{L}$ is better for large M. Hence, the surrogate linear model is often chosen as high dimensional, meaning that M is large with respect to n. The penalty term in the estimation criterion makes it possible to find a reasonable estimator. In the following we represent the principles of ridge and lasso.

6.4.1 Ridge Regression

The ridge regression was introduced in Hoerl and Kennard (1970). Here, the penalty term is the squared norm of the coefficients. In the theory of ill-posed problems, this method has been known as Tikhonov regularization since 1963. One of the interpretations of the name "ridge" is related to a linear model with correlated covariates. The density of a normal distribution with correlated components looks like a mountain with a ridge, see Figure 6.16.
Consider the linear regression model with intercept

$$\mathbf{Y} = \beta_0 \mathbf{1} + \mathbf{X}\beta + \varepsilon,$$

where $\mathbf{Y} = (y_1, \ldots, y_n)^T$, $\beta = (\beta_1, \ldots, \beta_M)^T$, $\mathbf{X} = (x_{ij})_{i=1,\ldots,n,\ j=1,\ldots,M}$, $\mathbf{1} = (1, \ldots, 1)^T$, β_0 is a scalar - the intercept. Then the ridge estimator is defined as

$$\widehat{\beta}_{ridge} = \arg\min_{\beta} \left(\sum_{i=1}^{n} \left(y_i - \beta_0 - \sum_{j=1}^{M} \beta_j x_{ij} \right)^2 + \lambda \sum_{j=1}^{M} (\beta_j)^2 \right), \quad (6.6)$$

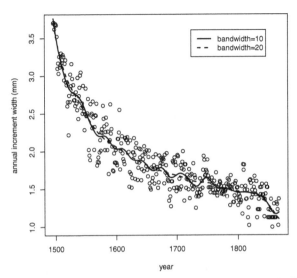

FIGURE 6.12: Methuselah data. Local quadratic regression with Gaussian kernel. Bandwidth 20 delivers oversmoothing; therefore, changes in the environmental conditions cannot be detected.

where λ is a tuning (smoothing) parameter that controls the penalty term. In contrast to the least squares estimator, the components of the ridge estimator are closer to each other.

The intercept can be estimated by

$$\widehat{\beta}_{0,ridge} = \overline{y} - \sum_{j=1}^{M} \overline{x}_j \widehat{\beta}_{j,ridge}.$$

Consider the linear regression model without intercept

$$\mathbf{Y} = \mathbf{X}\beta + \varepsilon. \tag{6.7}$$

The ridge regressor is defined by

$$\widehat{\beta}_{ridge} = \arg\min_{\beta} C_{ridge}(\beta, \lambda), \tag{6.8}$$

with

$$C_{ridge}(\beta, \lambda) = (\mathbf{Y} - \mathbf{X}\beta)^T (\mathbf{Y} - \mathbf{X}\beta) + \lambda \beta^T \beta.$$

The M-dim vector of the first derivatives of the criterion function is given by:

$$C_{ridge}^{\beta}(\beta, \lambda) = -2\mathbf{X}^T \mathbf{Y} + 2(\mathbf{X}^T \mathbf{X} + \lambda \mathbf{I})\beta.$$

Hence, we have

$$\widehat{\beta}_{ridge} = (\mathbf{X}^T \mathbf{X} + \lambda \mathbf{I})^{-1} \mathbf{X}^T \mathbf{Y} = \mathbf{Z}\widehat{\beta},$$

6.4 Classes of Restricted Estimators

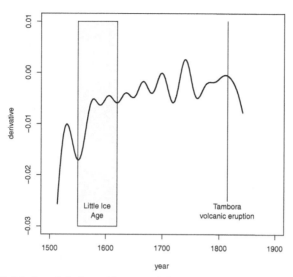

FIGURE 6.13: Methuselah data. Estimated first derivative based on the local cubic regression with bandwidth 15. Different climate conditions are visible.

where $\widehat{\beta} = \widehat{\beta}_{lse}$ is an arbitrary solution of

$$(\mathbf{X}^T\mathbf{X})\widehat{\beta} = \mathbf{X}^T\mathbf{Y},$$

and

$$\mathbf{Z} = (\mathbf{X}^T\mathbf{X} + \lambda \mathbf{I})^{-1}\mathbf{X}^T\mathbf{X}. \tag{6.9}$$

Note also that in case of $\det(\mathbf{X}^T\mathbf{X}) = 0$, the inverse $\mathbf{X}^T\mathbf{X} + \lambda \mathbf{I}$ exists and the ridge estimator is uniquely defined.

Let us discuss the variance-bias trade-off for the ridge estimator. Consider the mean squared error (MSE) under (6.7)

$$MSE(\widehat{\beta}_{ridge}) = E(\widehat{\beta}_{ridge} - \beta)^T(\widehat{\beta}_{ridge} - \beta) = V_{ridge}(\lambda) + B_{ridge}(\lambda),$$

with the variance term

$$V_{ridge}(\lambda) = tr Cov(\widehat{\beta}_{ridge}),$$

and the bias term

$$B_{ridge}(\lambda) = (\beta - E\widehat{\beta}_{ridge})^T(\beta - E\widehat{\beta}_{ridge}).$$

Using $Cov(\widehat{\beta}) = \sigma^2(\mathbf{X}^T\mathbf{X})^-$, - stands for the general inverse, the variance term can be calculated as follows

$$V_{ridge}(\lambda) = tr(\mathbf{Z}Cov(\widehat{\beta})\mathbf{Z}) = \sigma^2(\mathbf{X}^T\mathbf{X} + \lambda I)^{-1}\mathbf{X}^T\mathbf{X}(\mathbf{X}^T\mathbf{X} + \lambda I)^{-1}.$$

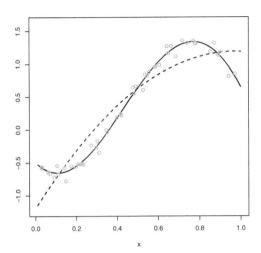

FIGURE 6.14: The connected line is the true underlying regression function of the gray observed points. The broken line is the best possible global quadratic approximation of the true regression curve.

Denote the eigenvalues of $\mathbf{X}^T\mathbf{X}$ by $\lambda_1, \ldots, \lambda_M$. We obtain

$$V_{ridge}(\lambda) = \sigma^2 \sum_{j=1}^{M} \frac{\lambda_j}{(\lambda_j + \lambda)^2}.$$

The variance term is monotone decreasing in λ. In order to calculate the bias term

$$B_{ridge}(\lambda) = \beta^T (\mathbf{I} - \mathbf{Z})^T (\mathbf{I} - \mathbf{Z}) \beta,$$

we have $\mathbf{I} - \mathbf{Z} =$

$I - (\mathbf{X}^T\mathbf{X} + \lambda\mathbf{I})^{-1}\mathbf{X}^T\mathbf{X} = (\mathbf{X}^T\mathbf{X} + \lambda\mathbf{I})^{-1}((\mathbf{X}^T\mathbf{X} + \lambda\mathbf{I}) - \mathbf{X}^T\mathbf{X}) = \lambda(\mathbf{X}^T\mathbf{X} + \lambda\mathbf{I})^{-1}$

Thus

$$B_{ridge}(\lambda) = \lambda^2 \beta^T (\mathbf{X}^T\mathbf{X} + \lambda\mathbf{I})^{-2} \beta.$$

The bias term is increasing in λ. Suppose an orthogonal design $\mathbf{X}^T\mathbf{X} = \mathbf{I}$, then

$$V_{ridge}(\lambda) = \sigma^2 \frac{M}{(1+\lambda)^2}, \quad B_{ridge}(\lambda) = \frac{\lambda^2}{(1+\lambda)^2} \beta^T \beta.$$

After summarizing, we obtain for some λ

$$MSE(\widehat{\beta}_{ridge}) = \frac{M}{(1+\lambda)^2} \sigma^2 + \frac{\lambda^2}{(1+\lambda)^2} \beta^T \beta \leq MSE(\widehat{\beta}).$$

This means that for some λ $\widehat{\beta}_{ridge}$ is better than BLUE $\widehat{\beta}$, see Figure 6.17.

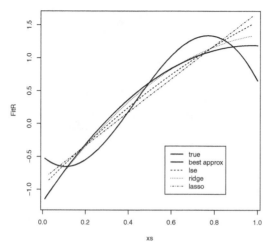

FIGURE 6.15: All three estimating procedures deliver a reasonable answer for the data in Figure 6.14.

6.4.2 Lasso

The name lasso is an acronym standing for "least absolute shrinkage and selection operator", introduced by Tibshirani (1996). Maybe they had in mind the illustration in Figure 6.18. The main idea is that a quadratic loss function on the error term is combined with an absolute loss on the parameter:

$$\|\beta\|_1 = \sum_{j=1}^{M} |\beta_j|,$$

$$\widehat{\beta}_{lasso} \in \arg\min_{\beta \in \mathbb{R}^M} \left(\|\mathbf{Y} - \mathbf{X}\beta\|^2 + \lambda \|\beta\|_1 \right).$$

The advantage of this method is that the problem can be solved by convex optimization. The lasso minimization problem has no explicit solution. The solution can be expressed as the solution of the KKT conditions (Karush-Kuhn-Tucker)

$$\mathbf{X}^T\mathbf{Y} - \mathbf{X}^T\mathbf{X}\beta = \lambda\gamma,$$

$$\gamma_i \in \begin{cases} sign(\beta_i) & \text{if } \beta_i \neq 0 \\ [-1, 1] & \text{if } \beta_i = 0. \end{cases}$$

The restricted optimization problems can be reformulated by the dual problems. For all λ in

$$\widehat{\beta}_{ridge} = \arg\min_{\beta \in \mathbb{R}^M} \left(\|\mathbf{Y} - \mathbf{X}\beta\|^2 + \lambda \|\beta\|^2 \right),$$

it exists a bound k_R such that

$$\widehat{\beta}_{ridge} = \arg\min \left(\|\mathbf{Y} - \mathbf{X}\beta\|^2 : \beta \in \mathbb{R}^M, \|\beta\|^2 \leq k_R \right).$$

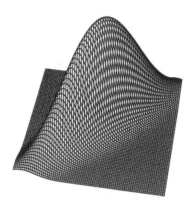

FIGURE 6.16: The density of a normal distribution with correlated components.

For all λ
$$\hat{\beta}_{lasso} \in \arg\min_{\beta \in \mathbb{R}^M} \left(\|\mathbf{Y} - \mathbf{X}\beta\|^2 + \lambda \|\beta\|_1 \right),$$

it exists a bound k_L such that

$$\hat{\beta}_{lasso} \in \arg\min \left(\|\mathbf{Y} - \mathbf{X}\beta\|^2 : \beta \in \mathbb{R}^M, \|\beta\|_1 \leq k_L \right).$$

In Figure 6.19, the dual problems are compared. The lasso method pushes the parameter to zero, Figure 6.20, which is why the lasso is an estimation method and a model choice method. The lasso solution delivers models with less basis functions.

The R code below presents the most important command lines of the artificial data set in Figure 6.14. The results are given in Figure 6.15.

R Code 6.4.33. Ridge, Lasso

```
### the example
x <- rbeta(50,1,1)
xs <- sort(x)
f1 <- xs^2-4*(xs-0.5)^3-0.5*xs^4-cos(4*xs)
y <- f1+rnorm(10,0,0.09)
### the global quadratic approximation
xx1 <- xx-mean(xx)
xx2 <- xx^2-mean(xx^2)
ff <- xx^2-4*(xx-0.5)^3-0.5*xx^4-cos(4*xx)
```

6.5 Spline Estimators

```
M <- lm(ff~xx1+xx2)
### lse
xs1 <- xs-mean(xs)
xs2 <- xs^2-mean(xs^2)
LSE <- lm(y~xs1+xs2)
### ridge
library(MASS)
RIDGE <- lm.ridge(y~xs1+xs2,lambda=seq(0,6,0.1))
### lasso
library(lars)
XX <- matrix(c(rep(1,50),xs1,xs2),ncol=3)
LASSO <- lars(XX,y,type="lasso",intercept=FALSE)
```

6.5 Spline Estimators

Spline estimators are based on spline approximations and linear regression methods. The word "spline" stands for pieces, in our case polynomials, that are put together smoothly. The x value, where the polynomial on the left side meets the other polynomial on the right side, is called a knot. The polynomials are chosen dependent of each other such that derivatives of the left polynomial at the knot coincide with the derivatives of the right polynomial. The turning parameters are: the number and positions of the knots, the degree of the polynomials, and smoothness of the composite function. In contrast to local polynomial methods, the partition of the x axis is determined in advance. The spline technique is a global approximation method.

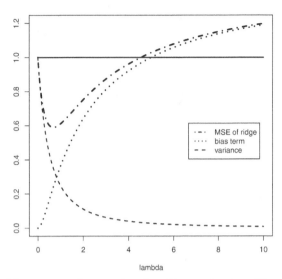

FIGURE 6.17: Mean squared error for the ridge estimator. There exist turning parameters λ which generate a smaller mean squared error than the least squares estimator.

6.5.1 Base Splines

Suppose that the underlying regression function $m(x)$ is approximated by a function $g(x)$, piecewise defined with no constraints at the boundary points:

$$g(x) = \begin{cases} k(x,\beta) & \text{for} \quad x \in [a,b) \\ h(x,\alpha) & \text{for} \quad x \in [b,c) \\ d(x,\eta) & \text{for} \quad x \in [c,d), \end{cases}$$

then

$$\arg\min_{\beta,\alpha,\eta} \sum (y_i - g(x_i))^2$$
$$= \arg\min_{\beta} \sum_{x_i \in [a,b)} (y_i - k(x_i,\beta))^2$$
$$+ \arg\min_{\alpha} \sum_{x_i \in [b,c)} (y_i - h(x_i,\alpha))^2$$
$$+ \arg\min_{\eta} \sum_{x_i \in [c,d)} (y_i - d(x_i,\eta))^2.$$

This means that the problem is divided in three independent regression problems. The situation is changed, if there are smoothness conditions on the boundary points b, c. For illustration, see Figure 6.22.

6.5 Spline Estimators

FIGURE 6.18: The lasso method picks up the leading horses.

Consider a piecewise linear continuous function:

$$g(x) = \begin{cases} \beta_1 + \beta_2 x & \text{for} \quad x \in [0, \xi_1) \\ \beta_3 + \beta_4 x & \text{for} \quad x \in [\xi_1, \xi_2) \\ \beta_4 + \beta_5 x & \text{for} \quad x \in [\xi_2, 1), \end{cases}$$

with the continuity conditions

$$\beta_1 + \beta_2 \xi_1 = \beta_3 + \beta_4 \xi_2, \quad \beta_3 + \beta_4 \xi_1 = \beta_5 + \beta_6 \xi_2 \,.$$

Only 4 free parameters are left. Fortunately, $g(x)$ can be presented as

$$g(x) = \sum_{j=1}^{4} \beta_j h_j(x),$$

with base functions

$$h_1(x) = 1, \; h_2(x) = x, \; h_3(x) = (x - \xi_1)_+, \; h_4(x) = (x - \xi_2)_+,$$

where

$$(x)_+ = \begin{cases} 0 & \text{for} \quad x < 0 \\ x & \text{for} \quad x \geq 0 \end{cases}.$$

Suppose $x_{k_1} < \xi_1 \leq x_{k_1+1}$ and $x_{k_2} < \xi_2 \leq x_{k_2+1}$. Then, the least squares minimization problem with constraints can be rewritten as a minimization problem related to the surrogate linear regression model $Y = \mathbf{H}\beta + \varepsilon$ where $\beta = (\beta_1, \beta_2, \beta_3, \beta_4)^T$ and the $n \times 4$ matrix $\mathbf{H}$ are defined as follows

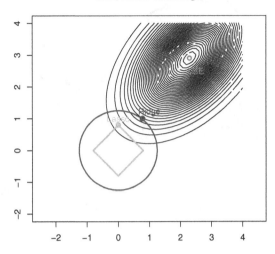

FIGURE 6.19: The contour lines respect the quadratic criteria function which should be minimized. The unconstraint minimum (LSE) lies in the center. The lasso estimate is located at a corner of the rhombus. The ridge estimator has circular constraints.

$$\mathbf{H} = \begin{pmatrix} 1 & x_1 & 0 & 0 \\ \vdots & \vdots & \vdots & \vdots \\ 1 & x_{k_1} & 0 & 0 \\ 1 & x_{k_1+1} & x_{k_1+1} - \xi_1 & 0 \\ \vdots & \vdots & \vdots & \vdots \\ 1 & x_{k_2} & x_{k_2+1} - \xi_1 & 0 \\ 1 & x_{k_2+1} & x_{k_2+1} - \xi_1 & x_{k_2+1} - \xi_2 \\ \vdots & \vdots & \vdots & \vdots \\ 1 & x_n & x_n - \xi_1 & x_n - \xi_2 \end{pmatrix}.$$

Definition: g is an order **M-spline** with knots $\xi_1 \ldots \xi_K$, where $\xi_1 < \ldots < \xi_K$ iff g is piecewise polynomial of order M-1 and has continuous derivatives up to order M-2.

For example, a piecewise continuous linear function is a $M = 2$ spline. A cubic spline is of order $M = 4$.

In order that g is a piecewise polynomial with continuous derivatives up to order $M - 2$, the values of the left polynomial and of the right polynomial

6.5 Spline Estimators

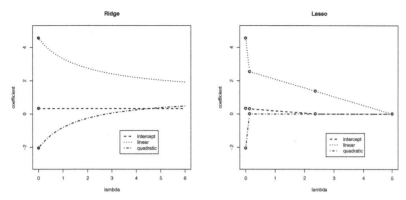

FIGURE 6.20: Left: Ridge estimators are smoothly shrinking by increasing λ. Right: Lasso estimators depend piecewise linearly on the tuning parameter. For large $\lambda > 5$, all estimated coefficients are equal to zero.

have to be the same
$$g_{\text{left}}(\xi) = g_{\text{right}}(\xi)$$
at each knot $\xi \in (a, b)$. In addition, the derivatives up to order $M - 2$ have to coincide
$$g_{\text{left}}^{(k)}(\xi) = g_{\text{right}}^{(k)}(\xi), \quad \text{for } k = 0, \ldots, M - 2,$$
see Figure 6.23. We have the following presentation of M-splines, which is beneficial for applying linear regression methods.

It holds: g is order M-spline with knots $\xi_1, \ldots, \xi_K$ iff
$$g(x) = \sum_{j=1}^{M+K} \beta_j h_j(x),$$
with
$$h_j(x) = x^{j-1}, \, j = 1, \ldots, M,$$
and
$$h_{M+l}(x) = (x - \xi_l)_+^{M-1}, \, l = 1, \ldots, K.$$
In Figure 6.21, some basis functions are plotted. Define $\mathbf{H} = (h_j(x_i))_{ij}$, we have to solve
$$\min_g \sum_{i=1}^{n}(y_i - g(x_i))^2 = \min_{\beta \in \mathbb{R}^{K+M}} \|Y - \mathbf{H}\beta\|^2. \tag{6.10}$$

The solution $\widehat{\beta}$ delivers the regression spline or base spline.
$$(\widehat{g}(x_i))_{i=1..n} = \mathbf{H}\widehat{\beta}, \quad \widehat{g}(x) = \sum_{j=1}^{M+K} \widehat{\beta}_j h_j(x).$$

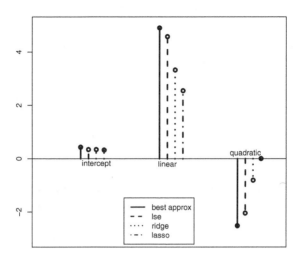

FIGURE 6.21: Comparison of the different estimates and the effect of shrinking.

Regression splines have often problems with the fit at the borders, which is why the following additional constraints are required: the spline should be linear on the border intervals $[a, \xi_1]$, $[\xi_K, b]$.

A **natural spline** is a cubic (order 4) spline which is linear on the border intervals $[a, \xi_1]$, $[\xi_K, b]$. In this case, we have a presentation which allows the application of linear regression methods, Figure 6.24.

It holds: g is a natural spline iff

$$g(x) = \sum_{j=1}^{K} \beta_j N_j(x),$$

with

$$N_1(x) = 1, \ N_2(x) = x, \ N_{k+2}(x) = d_k(x) - d_{K-1}(x), \ k = 1, \ldots, K-2$$

$$d_k(x) = \frac{(x - \xi_k)_+^3 - (x - \xi_K)_+^3}{\xi_K - \xi_k}.$$

The R package *splines* calculates the respective design matrices for the surrogate linear regression model, then the R code for linear regression is applied. The following R code is used for Figures 6.25 and 6.26.

6.5 Spline Estimators

R Code 6.5.34. Base spline, natural spline

```
x <- rbeta(100,0.3,0.5)
xs <- sort(x)
m <- sin(xs*10)+sin(xs*20)+(2*xs-1)^2 ### regression function
y <- m+rnorm(100,0,0.5)
plot(xs,y,"p")   ### (xs,y) simulated data
library(splines)
B <- bs(xs,knots=c(0.1,0.2,0.3,0.4,0.5,0.6,0.7,0.8,0.9))
Mb <- lm(y~B)
plot(xs,y,"p")
lines(xs,fitted(Mb),lwd=3,lty=2)
N <- ns(xs,knots=c(0.1,0.2,0.3,0.4,0.5,0.6,0.7,0.8,0.9))
Mn <- lm(y~N)
lines(xs,fitted(Mn),lwd=3,lty=2)
```

6.5.2 Smoothing Splines

Upon the first glance, smoothing splines are based on a quite different approach. We assume that the regression function belongs to the class of function

$$\mathcal{F} = \left\{ f : f : [a,b] \to \mathbb{R},\ f''\ continuous,\ \int f''(x)^2 dx < \infty \right\},$$

and start with a general least squares problem. We consider a least squares estimation criterion with penalty term $\int f''(x)^2 dx$:

$$RSS(f,\lambda) = \sum_{i=1}^{n} (y_i - f(x_i))^2 + \lambda \int f''(x)^2 dx.$$

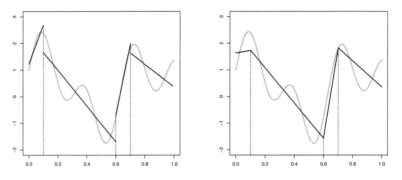

FIGURE 6.22: Left: Best piecewise linear approximation with knots at $0.1, 0.6, 0.7$. Right: Best linear spline approximation with knots at $0.1, 0.6, 0.7$.

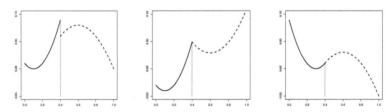

FIGURE 6.23: Left: No connection between the pieces. Center: Connected but not smooth. Right: Smooth connection.

The penalty term supports smooth functions with less fluctuations. Note that for $\lambda = 0$, we have $\min_{f \in \mathcal{F}} RSS(f, 0) = 0$, see Figure 6.3.
The solution $\widehat{f}_\lambda(x)$

$$\widehat{f}_\lambda(x) = \arg \min_{f \in \mathcal{F}} RSS(f, \lambda) \qquad (6.11)$$

is called a **smoothing spline**. The following theorem gives the connection between smoothing splines and natural splines.

Theorem 6.1 *The smoothing spline* $\widehat{f}_\lambda(x)$ *is a natural cubic spline with knots* $a < x_1 < \ldots < x_n < b$.

Proof:
Introduce $x_0 = a$, and $x_{n+1} = b$. Consider an arbitrary smooth function $\widetilde{g} \in \mathcal{F}$ with $\widetilde{g}(x_i) = z_i$, $i = 0, \ldots, n+1$.
Define a natural cubic spline with $g \in \mathcal{F}$ with knots $a < x_1 < \ldots < x_n < b$ and $g(x_i) = z_i$, $i = 0, 1, \ldots, n+1$, such that $\widetilde{g}(x_i) = g(x_i)$, for $i = 0, \ldots, n+1$.

6.5 Spline Estimators

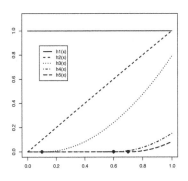
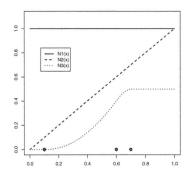

FIGURE 6.24: Left: Basis functions of a quadratic base spline with knots at $0.1, 0.6, 0.7$. Right: Basis functions of a natural spline with knots at $0.1, 0.6, 0.7$.

Then
$$RSS(\tilde{g}, \lambda) = \sum_{i=1}^{n} (y_i - z_i))^2 + \lambda \int \tilde{g}''(x)^2 dx,$$
and
$$RSS(g, \lambda) = \sum_{i=1}^{n} (y_i - z_i))^2 + \lambda \int g''(x)^2 dx.$$
For
$$RSS(\tilde{g}, \lambda) \geq RSS(g, \lambda),$$
it is enough to show
$$\int_a^b \tilde{g}''(x)^2 dx \geq \int_a^b g''(x)^2 dx,$$
and $\int \tilde{g}''(x)^2 dx > \int g''(x)^2 dx$ iff $\tilde{g}$ is no cubic spline.
Define $h(x) = \tilde{g}(x) - g(x)$, note $h(x_i) = 0$ for $i = 0, \ldots, n+1$.
Consider the integral $\int_a^b g''(x)h(x) dx$. Using the telescope expansion we get
$$\int_a^b g''(x)h''(x) dx = \sum_{i=0}^{n} \int_{x_i}^{x_{i+1}} g''(x)h''(x) dx,$$
and apply partial integration on each term
$$\int_{x_i}^{x_{i+1}} g''(x)h''(x) dx = g''(x)h'(x)\big|_{x_i}^{x_{i+1}} - \int_{x_i}^{x_{i+1}} g'''(x)h'(x) dx.$$
Note that g is a cubic spline, so on the intervals, $(x_i, x_{i+1}]$ g is a cubic function, implying that $g'''(x) = const$ and for $i = 1, \ldots, n-1$
$$\int_{x_i}^{x_{i+1}} g'''(x)h'(x) dx = const \int_{x_i}^{x_{i+1}} h'(x) dx = const \, (h(x_{i+1}) - h(x_i)) = 0.$$

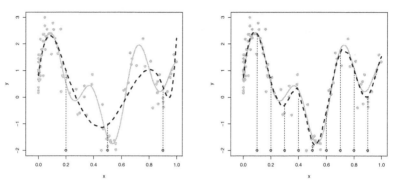

FIGURE 6.25: Left: Cubic base spline with a bad choice of knots. Right: Cubic base spline with a good choice of knots.

On the bonder intervals g is linear, implying that $g'''(x) = 0$ thus for $i = 0$ and $i = n$
$$\int_{x_i}^{x_{i+1}} g'''(x)h'(x)dx = 0.$$

Further

$$\sum_{i=0}^{n} g''(x)h'(x)\Big|_{x_i}^{x_{i+1}} = \sum_{i=0}^{n}(g''(x_{i+1})h'(x_{i+1}) - g''(x_i)h'(x_i)) = g''(a)h'(a) - g''(b)h'(b)$$

because g is natural cubic spline, that implies $g''(a) = g''(b) = 0$.
Thus
$$\int_a^b g''(x)h''(x)dx = 0,$$

and
$$\int_a^b g''(x)h''(x)dx = \int_a^b g''(x)\widetilde{g}''(x)dx - \int_a^b g''(x)^2 dx = 0.$$

Hence

$$0 \leq \int_a^b \left(g''(x) - \widetilde{g}''(x)\right)^2 dx = \int_a^b \widetilde{g}''(x)^2 dx - 2\int_a^b g''(x)\widetilde{g}''(x)dx + \int_a^b g''(x)^2 dx$$
$$= \int_a^b \widetilde{g}''(x)^2 dx - \int_a^b g''(x)^2 dx.$$

Thus
$$\int_a^b \widetilde{g}''(x)^2 dx \leq \int_a^b g''(x)^2 dx.$$

6.5 Spline Estimators

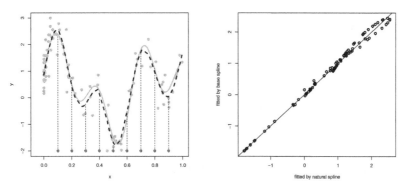

FIGURE 6.26: Left: Natural spline with good choice of knots. Right: In this example, the fitted values of the base spline and the natural spline are similar.

It remains to discuss the equality:

$$0 = \int_a^b (g''(x) - \widetilde{g}''(x))^2 \, dx$$

iff

$$g''(x) = \widetilde{g}''(x) \text{ for all } x.$$

Especially

$$g''(x) = \widetilde{g}''(x) \text{ for all } x \in [x_i, x_{i+1}).$$

On $[x_i, x_{i+1})$, $i = 1, \ldots, n-1$ g is cubic, thus $g''(x)$ is linear. We have

$$\alpha_i + \beta_i x = \widetilde{g}''(x), \text{ for all } x \in [x_i, x_{i+1}),$$

that implies

$$g(x) = \widetilde{g}(x) + c_o + c_1 x, \text{ for all } x \in [x_i, x_{i+1}],$$

but $g(x_i) = \widetilde{g}(x_i)$ and $g(x_{i+1}) = \widetilde{g}(x_{i+1})$ gives $c_0 = c_1 = 0$ and $g(x) = \widetilde{g}(x)$ for all x. For $i = 0$ and $i = n$ we have $g''(x) = \widetilde{g}''(x) = 0$.

□

This theorem provides the bridge between two concepts. The definition of the penalty term is the most important part. It also works with other criteria which are not least squares, but they have the same penalty term. Using this theorem we can solve the minimization problem (6.11). It holds

$$\min_{f \in \mathcal{F}} RSS(f, \lambda) = \min_{g \text{ natural spline}} RSS(g, \lambda).$$

The natural splines have a base presentation

$$g(x) = \sum_{j=1}^{n} \beta_j N_j(x), \quad g'(x) = \sum_{j=1}^{n} \beta_j N'_j(x), \quad g''(x) = \sum_{j=1}^{n} \beta_j N''_j(x),$$

and

$$\int_a^b g''(x)^2 dx = \sum_{j=1}^{n} \sum_{k=1}^{n} \beta_j \beta_k \int_a^b N''_j(x) N''_k(x) dx = \beta^T \Omega_n \beta,$$

with $\beta = (\beta_1, \ldots, \beta_n)^T$, $\Omega_n = \left(\int_a^b N''_j(x) N''_k(x) dx \right)_{jk}$.

The matrix notation for the sample is $Y = (y_1, \ldots, y_n)^T$, $\mathbf{N} = (N_j(x_i))_{ij}$. Then, the minimization problem in (6.11) can be rewritten as

$$\min_{g \text{ natural spline}} RSS(g, \lambda) = \min_{\beta} (Y - \mathbf{N}\beta)^T (Y - \mathbf{N}\beta) + \lambda \beta^T \Omega_n \beta.$$

This is a generalized ridge regression problem. The solution is given by

$$\widehat{\beta} = (\mathbf{N}^T \mathbf{N} + \lambda \Omega_n)^{-1} \mathbf{N}^T Y$$

and the smoothing spline $\widehat{f}_\lambda(x)$ is given by

$$\widehat{f}_\lambda(x) = \sum_{j=1}^{n} \widehat{\beta}_j N_j(x).$$

Particulary

$$(\widehat{f}_\lambda(x_i))_{i=1 \ldots n} = \mathbf{N}\widehat{\beta} = \mathbf{N}(\mathbf{N}^T \mathbf{N} + \lambda \Omega_n)^{-1} \mathbf{N}^T Y = S_\lambda Y.$$

$$S_\lambda = \mathbf{N}(\mathbf{N}^T \mathbf{N} + \lambda \Omega_n)^{-1} \mathbf{N}^T$$

is called the smoother matrix. In contrast to linear regression, S_λ is not a projection matrix. We have

$$\mathbf{N}^T \mathbf{N} + \lambda \Omega_n \succ \mathbf{N}^T \mathbf{N},$$

thus

$$\begin{aligned} S_\lambda S_\lambda &= \mathbf{N}(\mathbf{N}^T \mathbf{N} + \lambda \Omega_n)^{-1} \mathbf{N}^T \mathbf{N} (\mathbf{N}^T \mathbf{N} + \lambda \Omega_n)^{-1} \mathbf{N}^T \\ &\prec \mathbf{N}(\mathbf{N}^T \mathbf{N} + \lambda \Omega_n)^{-1} \mathbf{N}^T \mathbf{N} (\mathbf{N}^T \mathbf{N})^{-1} \mathbf{N}^T = S_\lambda. \end{aligned}$$

A projection matrix P is idempotent: $PP = P$, but $S_\lambda S_\lambda \prec S_\lambda$. This underlines the shrinking nature of S_λ. Analogously to the linear model, the trace $tr S_\lambda$ is called the effective degree of freedom and is used to determine the size of the tuning parameter λ.

The following R code is used to produce Figure 6.27.

R Code 6.5.35. Methuselah, smoothing spline, Example 6.2.

```
mm <- m[55:125] ### only Little Ice Age
plot(time[55:125],mm)
ssfit <- smooth.spline(time[55:125],mm,df=20)
xx <- seq(1550,1620,0.2)
lines(predict(ssfit,xx),lwd=2)
```

6.6 Wavelet Estimators

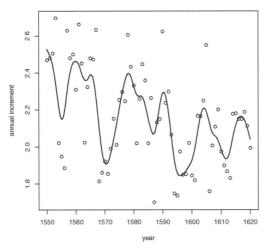

FIGURE 6.27: Methuselah data. Smoothing spline for studying the Little Ice Age 1550-1620.

6.6 Wavelet Estimators

6.6.1 Wavelet Base

Wavelet estimators can be considered as a special type of orthogonal series estimators. An orthogonal series estimator is constructed as follows. Suppose $m \in \mathcal{L}(e_1, \ldots, e_k, \ldots)$ such that $m(x) = \sum_{k=1} m_k e_k(x)$. The base functions $e_1, \ldots, e_k, \ldots$ form an orthogonal basis, this means that $\int e_k(x) e_j(x) w(x) dx = \delta_{kj}$. Then, an orthogonal series estimator is defined by

$$\widehat{m}_K(x) = \sum_{k=1}^{K} \widehat{m}_k e_k(x),$$

where

$$\widehat{m}_k = \frac{1}{n} \sum_{i=1}^{n} y_i \, e_k(x_i) \, w(x_i).$$

See Härdle (1989), Chapter 3.3.

The main trick of the construction of wavelet estimators is that we do not cut the series arbitrarily. The basis functions are grouped together and determine different orthogonal subspaces. Each subspace characterizes a new approximation step. Consider the complete orthogonal decomposition

$$\mathcal{L} = V_0 \oplus W_0 \oplus W_1 \ldots \oplus W_j \oplus \ldots.$$

The space V_0 is called reference space:
$$V_0 = \mathcal{L}\left(\{\Phi_{0,k}\}_{k\in\mathbb{Z}}\right).$$

The basis functions $\Phi_{0,k}$ are generated by translation (shifting) from the father function Φ:
$$\Phi_{0,k}(x) = \Phi(x-k).$$

For example, in the case of the Haar base, the father function is $I_{[0,1]}(x)$, such that $\Phi_{0,k} = I_{[k,k+1]}(x)$. The reference space V_0 includes all piecewise linear functions with jumps only at k, $k \in \mathbb{Z}$.
Generally it holds:
$$\int \Phi_{0,k}^2(x)dx = 1, \quad \int \Phi_{0,k}(x)\Phi_{0,l}(x)dx = 0, \text{ for } l \neq k,$$

thus $\{\Phi_{0,k}\}_{k\in\mathbb{Z}}$ is an orthogonal base system of V_0.
By rescaling (dilation), the base system $\{\Phi_{0,k}\}_{k\in\mathbb{Z}}$ can be extended. We define
$$\Phi_{j,k}(x) = 2^{\frac{j}{2}}\Phi_{j,0}(2^j x) = 2^{\frac{j}{2}}\Phi(2^j x - k).$$

For every fixed j, $\{\Phi_{j,k}\}_{k\in\mathbb{Z}}$ is an orthogonal base system:
$$\int \Phi_{j,k}^2(x)dx = 1, \quad \int \Phi_{j,k}(x)\Phi_{j,l}(x)dx = 0 \text{ for } l \neq k.$$

Define now
$$V_j = \mathcal{L}\left(\{\Phi_{j,k}\}_{k\in\mathbb{Z}}\right).$$

For Haar base, we have
$$\Phi_{j,k} = 2^{\frac{j}{2}} I_{[a,b]}(x), \quad \text{with } a = \frac{k}{2^j}, b = \frac{k+1}{2^j}.$$

In this case, the space V_j includes all piecewise linear functions with possible jumps only at $\frac{k}{2^j}$, $k \in \mathbb{Z}$. The distance between possible jumps cannot be larger than $\frac{1}{2^j}$.
Generally, given some father function, the construction is done such that
$$V_0 \subset V_1 \subset \ldots V_{j-1} \subset V_j \subset V_{j+1}\ldots.$$

For the complete decomposition of $\mathcal{L}$, we need to describe the functions which are not included in V_j but in V_{j+1}. Consider the orthogonal decomposition
$$V_{j+1} = V_j \oplus W_j.$$

The function which generates the base of W_0 is called the mother function, defined as
$$\psi(x) = \Phi(2x) - \Phi(2x-1), \text{ where } \int \psi(x)^2 = 1, \int \psi(x)\Phi(x) = 0.$$

6.6 Wavelet Estimators

Then, by translation and dilation we define

$$\psi_{j,k}(x) = 2^{\frac{j}{2}}\psi(2^j x - k),$$

where

$$\int \psi_{j,k}^2(x)dx = 1, \quad \int \psi_{j,k}(x)\psi_{j,l}(x)dx = 0, \text{ for } l \neq k,$$

and

$$\int \psi_{j,k}(x)\Phi_{j,l}(x)dx = 0, \text{ for all } l,k.$$

Thus, the $\{\psi_{j,k}\}_{k\in\mathbb{Z}}$ form an orthogonal base. We define the respective spaces by

$$W_j = \mathcal{L}\left(\{\psi_{j,k}\}_{k\in\mathbb{Z}}\right).$$

Thus, we have the orthogonal decomposition

$$V_J = V_0 \oplus W_0 \oplus W_1 \ldots \oplus W_{J-1}.$$

The mother function of the Haar basis is, Figure 6.28.

$$\psi(x) = \begin{cases} 1 & for \quad 0 \leq x < \frac{1}{2} \\ 0 & for \quad x = \frac{1}{2} \\ -1 & for \quad \frac{1}{2} < x \leq 1 \end{cases},$$

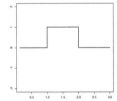 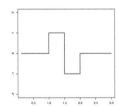 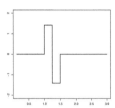

FIGURE 6.28: Left: Haar basis father function, Φ. Center: Mother function, Ψ. Right: Mother function, $\Psi_{1,0}$.

The Haar basis is often too coarse; however, many other wavelet bases have been created. One of the most common is the Daubechies symmlet-8 base which is now the default base in the R packet waveslim, see Figure 6.29. This base was invented by Ingrid Daubechies. She constructed different wavelet families. The father function Φ of symmlet-8 base has compact support and

$$\int_0^1 \Phi(x)dx = 1, \quad \int_0^1 x^j \Phi(x)dx = 0, \text{ for } j = 1,\ldots,8.$$

This implies that polynomials up to order 8 have a perfect wavelet presentation in V_0.

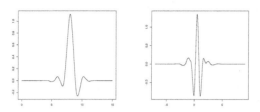

FIGURE 6.29: Left: Symmlet-8 basis father function. Right: Mother function.

6.6.2 Wavelet Smoothing

Assume sample with equidistant design points x_i and a sample size $n = 2^J$. In the case that the sample size cannot be given as $n = 2^J$, the trick is to expand the sample by artificial values. Then, the n-dimensional vector Y of observations can be completely decomposed by a discrete wavelet base. Define n-dimensional vectors

$$\Phi_{j,k} = (\Phi_{j,k}(x_i))_{i=1..n}, \quad \psi_{j,k} = (\psi_{j,k}(x_i))_{i=1..n},$$

and the related subspaces of the $\mathbb{R}^n$:

$$V_0 = \mathcal{L}\left(\{\Phi_{0,k}\}_{k \in \mathbb{Z}}\right), \ V_j = \mathcal{L}\left(\{\Phi_{j,k}\}_{k \in \mathbb{Z}}\right), \ W_j = \mathcal{L}\left(\{\psi_{j,k}\}_{k \in \mathbb{Z}}\right).$$

Then
$$\mathbb{R}^n = V_0 \oplus W_0 \oplus W_1 \oplus \ldots \oplus W_{J-1}.$$

Especially for $x_i = \frac{i}{n}$ we have $V_0 = \mathcal{L}(\{\Phi_{0,1}\})$ and

$$\dim V_0 = 1, \ \dim W_0 = 1, \ \dim W_1 = 2, \ldots, \dim W_j = 2^{j-1},$$

such that
$$\mathbb{R}^n = V_0 \oplus W_0 \oplus W_1 \oplus \ldots \oplus W_{J-1},$$

and the dimension of all subspaces are added to n:

$$1 + 1 + 2 + 4 + \ldots + 2^{j-1} + \ldots + 2^{J-1} = 2^J = n.$$

Hence, all observations can be presented as

$$Y = W\theta,$$

with $\theta \in \mathbb{R}^n$ and the $n \times n$ matrix

$$W = (\Phi_{0,1}, \psi_{0,1}, \psi_{1,1}, \psi_{1,2}, \ldots \psi_{J-1,J-1}).$$

The parameter θ can be estimated by least squares or shrinking methods. Note that because of the orthogonality it holds: $W^T W = I_n$ and

$$\widehat{\theta}_{lse} = (W^T W)^{-1} W^T Y = W^T Y = \arg\min \|Y - W\theta\|^2.$$

6.6 Wavelet Estimators

$\widehat{\theta}_{lse}$ is called wavelet transform. Further

$$\widehat{Y} = W\widehat{\theta}_{lse} = Y = P_{V_0}Y + P_{W_0}Y + \ldots + P_{W_{J-1}}Y,$$

where $P_{W_j}Y$ denotes the projection of Y on the subspace W_j.
The wavelet estimator is given by

$$\widehat{Y}_{wave} = P_{V_0}Y + P_{W_0}Y + \ldots + P_{W_{k-1}}Y = P_{V_k}Y.$$

The remainder terms

$$P_{W_k}Y + \ldots + P_{W_{J-1}}Y$$

are interpreted as noise or error terms. The choice of k controls the degree of smoothing (depth of decomposition).
Wavelet shrinkage is given by

$$\widehat{\theta}_{lasso} = \arg\min \|Y - W\theta\|^2 + \lambda \|\theta\|_1.$$

The following R code is used to produce Figure 6.30 and Figure 6.31.

R Code 6.6.36. Methuselah, wavelet, Example 6.2.

```
library(waveslim)
### take only 64 years 1555-1618, J=6
mm <- m[60:123]
W <- dwt(mm,n.levels=2)
plot(1:32,W$d1, ylim=c(-0.5,5),"h")
lines(1:16+0.2,W$d2,"h")
lines(1:16+0.4,W$s2,"h")
### wavelet thresholding for the transforms
W.sure <- sure.thresh(W,max.level=2,hard=TRUE)
plot(1:32,W.sure$d1, ylim=c(-0.5,5),"h")
```

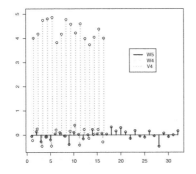
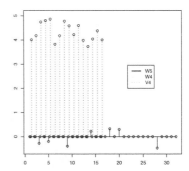

FIGURE 6.30: Methuselah data 1555-1619. Left: Wavelet transform. Right: Wavelet transform after using lasso.

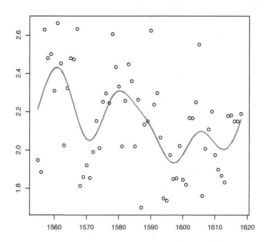

FIGURE 6.31: Methuselah data during the period of the Little Ice Age. In this case, the wavelet estimates are the same for the usual least squares method and lasso shrinkage.

```
points(1:16+0.2,W.sure$d2,"h")
lines(1:16+0.4,W.sure$s2,"h")
### comparison smoothing with sure and with out
plot(time[60:123],mm,"p",xlab="",ylab="")
WA <- mra(mm,J=3)
lines(time[60:123],WA$S3)
WA.sure <- sure.thresh(WA,max.level=2,hard=TRUE)
lines(time[60:123],W1A.sure$S3)
```

The following R code is used to produce Figure 6.32.

R Code 6.6.37. Methuselah, wavelet

```
library(waveslim)
W2 <- mra(m.exp,wf="la8",J=3,boundary="periodic")
par(mfcol=c(5,1),pty="m",mar=c(5-2,4,4-2,2))
mts <- ts(m,start=c(1496))
plot(mts,ylab="annual increments",xlab="year")
V5 <- ts(W2[[4]][1:372],start=c(1496))
W6 <- ts(W2[[3]][1:372],start=c(1496))
W7 <- ts(W2[[2]][1:372],start=c(1496))
W8 <- ts(W2[[1]][1:372],start=c(1496))
plot(V5)
plot(W6)
plot(W7)
plot(W8)
```

6.7 Choosing the Smoothing Parameter

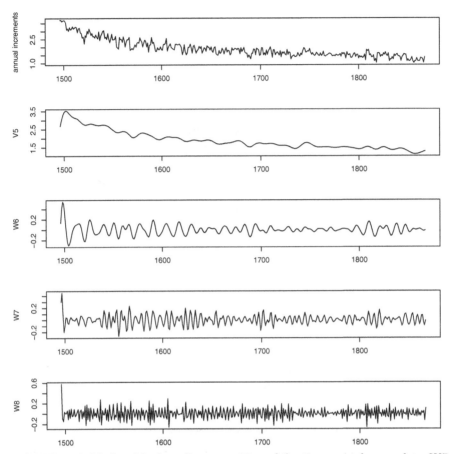

FIGURE 6.32: Methuselah data. Decomposition of the time series by wavelets. $W7$ shows a different shape under the Little Ice Age (1550-1620).

6.7 Choosing the Smoothing Parameter

All proposed estimators in this section depend on a smoothing parameter. The kernel estimators have the bandwidth h, the smooth spline estimators have the smoothing parameter λ, the base spline estimators have the number and position of the knots, and the wavelets have k (depth of decomposition).

General principle:

The aim is to find $m(x)$.

Start with a class of estimators $\widehat{m}_h$ depending on an additional parameter h.

1. Choose a starting value h_0 for h.
2. Calculate $\widehat{m}_{h_0}(x)$.

FIGURE 6.33: Principle of an iterative algorithm.

3. Estimate the criterion function of your interest $C(h)$ (for instance MISE, BIAS, MSE, AMISE) by $\widehat{C}(h)$ for each h.
4. Choose the new parameter: $h_1 := \arg\min \widehat{C}(h)$.
5. Take h_1 as the new starting value, see Figure 6.33.

Note that in some cases (like for cross-validation method) the estimator $\widehat{C}(h)$ does not depend on $\widehat{f}_{h_0}(x)$, meaning that there is no iteration in the estimation step. For more details, see Härdle (1989), Chapter 5.

6.8 Bootstrap in Regression

This section is a continuation to Chapter 3, Section 3.5 on bootstrap in regression.

General principle

1. Generate bootstrap samples from the data. This can be done by:
    - A parametric bootstrap

    $y_{ij}^* = \widehat{m}(x_i) + \varepsilon_{ij}^*, \quad j = 1\ldots B, i = 1,\ldots,n, \; \varepsilon_{ij}^*$ generated by P.

6.8 Bootstrap in Regression

- A nonparametric bootstrap

$$y_{ij}^* = \widehat{m}(x_i) + \varepsilon_{ij}^*, \quad j = 1, \ldots, B, \ i = 1, \ldots, n,$$

$$\varepsilon_{ij}^* \text{ sampled from } e_i = y_i - \widehat{m}(x_i) - \frac{1}{n}\sum_{i=1}^{n}(y_i - \widehat{m}(x_i)).$$

- In the case of random X variables, we can apply a case-wise bootstrap. (y_j^*, x_j^*) are sampled from the data.

2. From each bootstrap sample, the estimator of m or of derivatives of m are calculated.
3. A confidence band is calculated by percentile bootstrap intervals at each x (basic bootstrap intervals).
4. Perform the bootstrap test with the help of the confidence bands.

This was applied to the Methuselah data, see Figure 6.34.

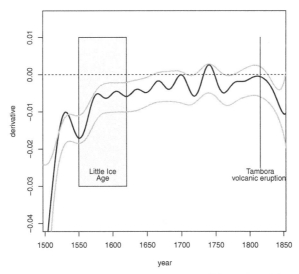

FIGURE 6.34: Methuselah data. 0.95-bootstrap confidence band for the first derivative.

We use the following steps for the Methuselah data:

1. The regression function is estimated by a smoothing spline with df=20 for the whole time series. Wavelets are not applied because the sample size is not 2^J. Nadaraya Watson estimators are not applicable because the data set forms a time series with fixed equidistant

x-values. The residuals are calculated and centered. The Shapiro test gives a p-value= 0.2316. This means that we cannot reject normal distribution for the residuals.

2. $B = 100$ bootstrap samples are generated using parametric bootstrap. The bootstrap errors are generated from a normal distribution with expectation zero and the estimated variance of the residuals. Then they are added to the smoothing spline estimates.

3. For each bootstrap sample, the derivative is estimated using a local cubic polynomial with the bandwidth 15.

4. For each time point, the percentile bootstrap interval is calculated for the derivative. These bounds are coarse that is why the bounds are smoothed by a smoothing spline.

Note that the whole procedure depends essentially on the choice of the smoothing parameters. Here, we apply an empirical choice just proving different variants.

R Code 6.8.38. Methuselah data, bootstrap confidence band, Example 6.2.

```
ssfit <- smooth.spline(time,m,df=20)
resid <- m-ssfit$y
shapiro.test(e)
sd(e)
library(KernSmooth)
LQ1 <- locpoly(time,m,degree=3,bandwidth=15,drv=1)
tq <- LQ1$x
L <- length(tq)
mq <- LQ1$y
plot(tq,mq,"l",xlim=c(1510,1840),ylim=c(-0.04,0.015))
### bootstrap confidence
B <- 100
YYD <- matrix(nr=372,nc=B)
FYD <- matrix(L,nc=B)
SD <- matrix(L,nc=B)
for(b in 1:B){
    YY[,b] <- ssfit$y+rnorm(372,0,sd(e))
    FYD[,b] <- locpoly(time,YY[,b],degree=3,bandwidth=15,drv=1)$y}
for( i in L){
    SD[i,] <- sort(FYD[i,])}
ssfit.1 <- smooth.spline(tq,SD[,0.025*B],df=20)
ssfit.2 <- smooth.spline(tq,SD[,0.925*B],df=20)
lines(tq,ssfit.1$y)
lines(tq,ssfit.2$y)
```

6.9 Problems

1. Analyze the data set of "longley" (data("Longley")).
   (a) Give a short explanation of the data.
   (b) Calculate a linear regression.
   (c) Calculate a ridge regression with different lambda.
   (d) Try to interpret the different estimators.
2. Generate your own data set.
   (a) Create an interesting "true" regression curve (for example with two modes, or with a jump).
   (b) Generate input variables with two different distributions (for example: uniform, truncated normal, beta).
   (c) Generate the outputs for these two different input data sets. Now you have two different data sets.
   (d) Fit base splines and natural splines to both samples. Choose different positions for knots and also use a different number of knots.
   (e) Give one example of a good fit. Explain why it is good.
   (f) Give one example of a bad fit due to overfitting.
   (g) Give one example for a bad fit due to an inconvenient choice of knots.
   (h) Fit a smoothing spline with an automatically chosen smoothing parameter.
3. (a) Generate a time series (equidistant "time points") with $N = 2^J$, for which the wavelet procedure with "Haar" delivers a good estimation and a usual spline method "oversmoothes" the shape of the curve.
   (b) Generate another time series (equidistant "time points") with $N = 2^J$, for which the wavelet procedure with "Haar" delivers a bad estimation and a usual spline method gives a good approximation.
   (c) Apply to both simulated data sets a wavelet approximation with a symmlet-8 base. Compare the approximations with the results in (a) and (b).
4. Consider the toy sample

$x * N$	1	2	3	4	5	6	7	8
y	$10+\sqrt{2}$	$10+\sqrt{2}$	$10-\sqrt{2}$	$10-\sqrt{2}$	10	10	12	8

   (a) Consider the Haar basis. Plot the basis functions for all subspaces $V_0, W_0, \ldots$.

(b) Determine the decomposition of y.

(c) Write the wavelet transform.

5. Given the following data set

x	-3	-2	-1.5	-1	-0.5	0	0.5	1	1.5	2	3
y	9.83	6.30	6.95	4.23	1.78	1.70	1.96	6.38	9.18	13.18	18.04

(a) Create a scatter plot.

(b) Calculate an estimator $\widehat{f}$ at the design points x_i, $i = 1, \ldots, n$ with

$$\widehat{f} \in \arg\min_{f \in \mathcal{F}} \sum_{i=1}^{n} (y_i - f(x_i))^2,$$

where $\mathcal{F}$ denotes the set of all continuous functions defined on $[-3, 3]$.

(c) Plot this estimator $\widehat{f}$ in the scatter plot.

(d) Calculate the estimator $\widetilde{f}(x)$ for $x \in [-3, 3]$ with

$$\widetilde{f}_1 \in \arg\min_{f \in \mathcal{F}_0} \sum_{i=1}^{n} (y_i - f(x_i))^2,$$

where $\mathcal{F}_0$ denotes the set of all continuous functions which are linear on $[-3, 0]$ and linear on $[0, 3]$ and with $f(0) = 1$.

(e) Plot this estimator $\widehat{f}_1$ in the scatter plot.

(f) Is this estimator $\widehat{f}_1$ a linear spline function?

6. Given the following data set

x	-1	-0.5	0	0	0.5	1
z	0	0	-1	1	0	0
y	-2.39	-0.68	-2.55	0.82	0.6	2.23

related to the model

$$y_i = \beta_1 x_i + \beta_2 z_i + \varepsilon_i, \quad E\varepsilon_i = 0.$$

(a) Calculate the ridge estimator for $\beta^T = (\beta_1, \beta_2)$ for $\lambda = 0.5$.

(b) Calculate the ridge estimator for $\beta^T = (\beta_1, \beta_2)$ for $\lambda = 0.2$.

(c) Calculate the cross-validation criteria for both estimators.

(d) Which ridge estimator do you prefer?

6.9 Problems

7. Given the following data set

x	-2	-1.5	-1	0	0.5	1	1.5	2
y	5.51	2.21	1.31	-1.1	0.30	2.46	4.31	6.07.

   Consider the estimator

   $$\widehat{f} \in \arg\min_{f \in \mathcal{F}} \sum_{i=1}^{n} (y_i - f(x_i))^2,$$

   where $\mathcal{F}$ denotes the set of all quadratic base splines with knots at $\xi_1 = -1.5, \xi_2 = 1$.

   (a) Which properties have a quadratic base spline with knots at $\xi_1 = -1.5, \xi_2 = 1$?
   (b) Plot base functions.
   (c) Formulate the related linear model.
   (d) Calculate the fitted spline.

8. Given a training $(x_i, y_i)_{i=1,..,n} = (X, Y)$, with $Var(y) = 1$. The design points are $uniform$ -distributed on $[0, 1]$. Consider the simplified kernel estimator for the regression function,

   $$\widehat{f}(z) = \frac{1}{nh} \sum_{i=1}^{n} K\left(\frac{z - x_i}{h}\right) y_i,$$

   where K is a given kernel with

   $$\int K(s)ds = 1, \quad \int K(s)sds = 0, \quad \int K(s)^2 ds = 0.5, \quad \int s^2 K(s)ds = 0.8.$$

   (a) Calculate the leading term of the bias term $E\widehat{f}(x)$.
   (b) Calculate the leading term of the variance $Var\widehat{f}(x)$.
   (c) Calculate the leading term of the bias term $(E\widehat{f}(x) - f(x))^2$.
   (d) Give a recommendation for an adaptive asymptotic bandwidth at $z = 0.6$, if the preliminary estimator is $\widetilde{f}(0.6) = 3$; $\widetilde{f}'(0.6) = 1$, $\widetilde{f}''(0.6) = 0.1$.
   (e) Using this bandwidth, which rate has the MSE at $z = 0.6$?

References

M. Abramowitz and I. A. Stegun. *Handbook of Mathematical Functions*. Dover, New York, 1964.

Y. F. Atchade and J. S. Rosenthal. On adaptive Markov Chain Monte Carlo Algorithms. *Bernoulli*, 11(5):815–828, 2005.

C.B. Bell, D. Blackwell, and L. Breiman. On the completeness of order statistics. *The Annals of Mathematical Statistics*, 31(794-979), 1960.

Y. K. Belyaev. *Bootstrap, resampling and Mallows metric*. Lecture Notes, Umeå University, 1995.

R. Beran. Prepivoting to reduce level error of confidence sets. *Biometrika*, 74 (3):457–468, 1987.

R. Beran. Prepivoting test statistics: A bootstrap view of asymptotic refinements. *Journal of the American Statistical Assosiation*, 83(403):687–697, 1988.

P. J. Bickel, F. Götze, and W. R. van Zwet. Resampling fewer than n observations: gains, losses, and remedy for losses. *Statistica Sinica*, 7(1):1–31, 1997.

G. Bloom. *Sannolikhetsteori med tillämpningar*. Studentliteratur, Lund, 1984.

G. E. P. Box and M. E. Muller. A note on the generation of random normal deviates. *Ann. Math. Statist.*, 29:610–611, 1958.

B. Efron, T. Hastie, I. Johnstone, and R. Tibshirani. Least angle regression. *The Annals of Statistics*, 32(2):407–499, 2003.

B. M. Brown, P. Hall, and G. A. Young. The smoothed median and the bootstrap. *Biometrika*, 88:519–534, 2001.

J. R. Cook and L. A. Stefanski. Simulation-Extrapolation Estimation in parametric measurement error models. *JASA*, 89:1314–1327, 1994.

A. C. Davison and D. V. Hinkley. *Bootstrap Methods and their Application*. Cambridge University Press, 1997.

A. P. Dempster, N. M. Laird, and D. B. Rubin. Maximum likelihood from incomplete data via EM algorithm. (with discussion). *J. Roy. Stat. Soc. B*, 39:1–38, 1977.

L. Devroye. *Non Uniform Random Variate Generation*. Springer, 1985.

T. J. DiCicco and B. Efron. Bootstrap confidence intervals. *Statistical Science*, 11(3):189–228, 1996.

B. Efron. Bootstrap methods: another look at the jackknife. *Ann. Statist.*, 7: 1–26, 1979.

B. Efron. Better bootstrap confidence intervals. *JASA*, 82:171–185, 1987.

B. Efron and R. J. Tibshirani. *An Introduction to the Bootstrap*, Chapman and Hall, 1993.

M. Eklund and S. Zwanzig. Simsel: a new simulation method for variable selection. *J Comp and Sim*, 85(515-527), 2011.

D. A. Freedman. Bootstrapping regression models. *Ann. Statist.*, pages 1218–1228, 1981.

W. A. Fuller. *Measurement Errors Models*. Wiley, 1987.

T. Gasser, H-G Müller, and V. Mammitzsch. Kernels for nonparametric curve estimation. *JRSS. B*, 47(2):282–252, 1985.

S. Geman and D. Geman. Stochastic relocation, Gibbs distributions, and the Bayesian restoration of images. *IEEE Transaction on Pattern Analysis and Machine Intelligence*, 6(6):721–741, 1984.

G. H. Givens and J. A. Hoeting. *Computational Statistics*. Wiley, 2005.

A. Gut. *An Intermediate Course in Probability*. Springer-Verlag, 1991.

P. Hall. On convergence rates in nonparametric problems. *Int. Stat.Rev.*, 57: 45–58, 1989.

W. Härdle. *Applied Nonparametric Regression*. Cambridge, 1989.

W. Härdle, J. Horowitz, and J.-P. Kreiss. Bootstrap methods for time series. *International Statistical Review*, 71(2):435–459, 2003. ISSN 03067734.

W.K. Hastings. Monte Carlo sampling methods using Markov chains and their applications. *Biometrika*, 57:97–109, 1970.

A. E. Hoerl and R. W. Kennard. Ridge regression biased estimation for nonorthogonal problems. *Technometrics*, 12(1), 1970.

References

C. Jentsch and J-P. Kreiss. The multiple hybrid bootstrap – resampling multivariate linear processes. *Journal of Multivariate Analysis*, 101:2320–2345, 2010.

W. J. Kennedy and J. E. Gentle. *Statistical Computing*, volume 33 of *Statistics: Textbooks and Monographs*. Marcel Dekker, 1980.

H. R. Künsch. The jackknife and the bootstrap for general stationary observations. *The Annals of Statistics*, 17(3):1217–1241, 1989.

K. Lange. *Numerical Analysis for Statisticians*. Springer, 1999.

H. Liero and S. Zwanzig. *Introduction to the Theory of Statistical Inference*. Chapman & Hall/CRC, 2011.

S. J. Lui. *Monte Carlo Strategies in Scientific Computing*. Springer Series in Statistics. Springer, 2001. ISBN 0-387-95230-6.

E. Mammen. Bootstrap and wild bootstrap for high dimensional linear models. *The Annals of Statistics*, 21(1):255–285, 1993.

P. Marjoram, J. Molitor, V. Palgnol, and S. Tavare. Markov chain Monte Carlo without likelihoods. *Proceedings of the National Academy of Sciences of the United States of America*, 100(26):15324–15328, 2003.

G. Marsaglia and A. Zaman. Technical Report. *The KISS generator*. Department of Statistics, Florida State University, Tallahassee, Florida, 1993.

N. Metropolis, A. Rosenbluth, M. N. Rosenbluth, A. H. Teller, and E. Teller. Equations of state calculations by fast computing machines. *Journal of Chemical Physics*, 21(6):1087–1091, 1953.

A. Miller. *Subset Selection in Regression*. Chapman & Hall CRC, 2002.

D. N. Politis and P. J. Romano. The stationary bootstrap. *JASA*, 89(428): 1303–1313, 1994.

J. Polzehl and S. Zwanzig. On a symmetrized simulation extrapolation estimator in linear errors-in-variables models. *Computational Statistics and Data Analysis*, 47(675-688), 2003.

J. G. Propp and D. B. Wilson. Random Structures & Algorithms. 9(1–2): 223–252, 1996.

B. D. Ripley. *Stochastic Simulation*. Wiley, 1987.

C. P. Robert and G. Casella. *Monte Carlo Statistical Methods*. Springer Texts in Statistics. Springer, 1999. ISBN 0-387-98707-X.

M. Rosenblatt. Remarks on some nonparametric estimates of a density function. *Ann. Math. Statist.*, 27(3):832–837, 1956.

D. B. Rubin. *Multiple Imputation for Nonresponse in Surveys.* Wiley-Interscience, 2004.

B. R. Schöne, J. Fiebig, M. Pfeiffer, R. Gleß, J. Hickson, A.L.A. Johnson, W. Dreyer, and Oschmann W. Climate records from a bivalved Methuselah, (*Arctica islandica*, mollusca; iceland). *Palaeogeography Palaeoclimatology Palaeoecology*, 228(130-148), 2005.

S.J. Sheather and M-C. Jones. A reliable data-based bandwidth selection method for kernel density estimation. *Journal of Royal Statistical Society*, 53(683-690), 1991.

K. Singh. On the asymptotic accuracy of Efron's bootstrap. *The Annals of Statistics*, 9(6):1187–1195, 1981.

R. Tibshirani. Regression shrinkage and selection via the lasso. *Journal of the Royal Statistical Society. Series B (Methodological)*, 58:267–288, 1996.

G. C. G. Wei and M. A. Tanner. A Monte Carlo implementation of the EM algorithm and the poor man's data augmentation algorithms. *Journal of the American Statistical Association*, 85(411), 1990.

N. A. Weiss. *Introductory Statistics.* Pearson, 2015.

Y. Wu, D. D. Boos, and L. A. Stefanski. Controlling variable selection by the addition of pseudovariables. *Journal of the American Statistical Association.*, 102, 2007.

A. Zygmund. *Trigonometric Series.* Cambridge University Press, 2003.

Index

Accept-reject method, 16
Adapted random-walk Metropolis, 47
Adaptive MCMC, 46
AIC-Forward, 130
Approximate Bayesian computation, ABC, 52
Approximate Bayesian computation, ABC-MCMC, 54

Bandwidth selection, 154
Base splines, 182
Blockwise Bootstrapping, 98
Bootstrap, 61
Bootstrap and Monte Carlo, 68
Bootstrap confidence sets, 75
Bootstrap distribution, 61
Bootstrap For Time Series, 97
Bootstrap hypothesis tests, 86
Bootstrap in Regression, 91, 200
Bootstrap Pivotal Methods, 78
Bootstrap replication, 61
Bootstrap samples, 61, 63
Box-Muller Algorithm, 14
Burn-in, 38, 47

Calibration, 116
Congruential generators, 5

Density Estimation, 141

Efron's Bootstrap, 64, 65
EM Algorithm, 106
EM Algorithm: Mixture Distributions, 112
Envelope Accept–Reject methods, 20
Errors-in-variables models, 118

F-Backward procedure, 124
F-Forward procedure, 124
FSR-Forward procedure, 130

Generalized inverse function, 1
Gibbs sampler, 48

Histogram, 143
Hit and run, 44
Hit or miss, 28

Importance sampling, 30
Independent MC, 27
Instrumental density, 16
Inverse method, 1

Kernel density estimator, 145, 147
Kernel regression smoothing, 166
KISS generator, 8

Langevin Metropolis Hastings, 45
Lasso, 179
Local regression, 169

M-spline, 184
Markov Chain Monte Carlo, 35
Markov Chain Monte Carlo, MCMC, 37
MCEM Algorithm, 114
Metropolis-Hastings Algorithm, 38, 39
Metropolized Independence Sampler, 43
Minimax, 158
Mixed-congruential, 5
Monte Carlo Method, 26
Monte Carlo tests, 87

Moon bootstrap, 64
Multiple imputations, 115
Multiple-try Metropolis-Hastings, 46
Multiplicative-congruential, 5
Nadaraya-Watson estimation, 168

Natural spline, 186
Nearest neighbor estimator, 157
Nonparametric (smoothed)
    bootstrap, 64
Nonparametric delta method, 79
Nonparametric regression, 163

Old Faithful, 153
Orthogonal series estimator, 157

Parametric bootstrap, 64
Parzen estimator, 147
Perfect Simulation, 47
Period, 5
Pivotal method, 76
Prepivoting Confidence Set, 83
Pseudo-random number generator, 4

Quantile function, 1

Random-Scan Gibbs Sampler, 48
Random-walk Metropolis, 41
RANDU, 6
Regression, 163

Rejection Sampling, 16
Restricted estimators, 173
Ridge regression, 175

Seed, 5
Shift-register generator, 8
Shift-register technique, 8
SIMEX, 115, 117, 119
SimSel, 132
Smoothing parameter, 199
Smoothing splines, 187
Spline estimators, 181
Squeezing Algorithm, 20
Stationary Bootstrap, 99
SYMEX, 121
Systematic-Scan Gibbs Sampler,
    49

Target density, 16
Transformation methods, 11
Trial distribution, 16

Variable selection, 123
Variance reduction, 28, 29
Violin plot, 33

Wavelet base, 193
Wavelet estimators, 193
Wavelet smoothing, 196
Wild bootstrap, 64